RESPONDING TO CAPABILITY SURPRISE

A Strategy for U.S. Naval Forces

Committee on Capability Surprise on U.S. Naval Forces

Naval Studies Board

Division on Engineering and Physical Sciences

NATIONAL RESEARCH COUNCIL
OF THE NATIONAL ACADEMIES

THE NATIONAL ACADEMIES PRESS
Washington, D.C.
www.nap.edu

THE NATIONAL ACADEMIES PRESS 500 Fifth Street, NW Washington, DC 20001

NOTICE: The project that is the subject of this report was approved by the Governing Board of the National Research Council, whose members are drawn from the councils of the National Academy of Sciences, the National Academy of Engineering, and the Institute of Medicine. The members of the committee responsible for the report were chosen for their special competences and with regard for appropriate balance.

This study was supported by Contract No. N00014-10-G-0589, DO #5 between the National Academy of Sciences and the Department of the Navy. Any opinions, findings, conclusions, or recommendations expressed in this publication are those of the author(s) and do not necessarily reflect the views of the organizations or agencies that provided support for the project.

International Standard Book Number-13: 978-0-309-27837-9
International Standard Book Number-10: 0-309-27837-6

Copies of this report are available free of charge from:

Naval Studies Board
National Research Council
The Keck Center of the National Academies
500 Fifth Street, NW
Washington, DC 20001

Additional copies of this report are available from the National Academies Press, 500 Fifth Street, NW, Keck 360, Washington, DC 20001; (800) 624-6242 or (202) 334-3313; http://www.nap.edu.

Printed in the United States of America

THE NATIONAL ACADEMIES

Advisers to the Nation on Science, Engineering, and Medicine

The **National Academy of Sciences** is a private, nonprofit, self-perpetuating society of distinguished scholars engaged in scientific and engineering research, dedicated to the furtherance of science and technology and to their use for the general welfare. Upon the authority of the charter granted to it by the Congress in 1863, the Academy has a mandate that requires it to advise the federal government on scientific and technical matters. Dr. Ralph J. Cicerone is president of the National Academy of Sciences.

The **National Academy of Engineering** was established in 1964, under the charter of the National Academy of Sciences, as a parallel organization of outstanding engineers. It is autonomous in its administration and in the selection of its members, sharing with the National Academy of Sciences the responsibility for advising the federal government. The National Academy of Engineering also sponsors engineering programs aimed at meeting national needs, encourages education and research, and recognizes the superior achievements of engineers. Dr. C. D. Mote, Jr., is president of the National Academy of Engineering.

The **Institute of Medicine** was established in 1970 by the National Academy of Sciences to secure the services of eminent members of appropriate professions in the examination of policy matters pertaining to the health of the public. The Institute acts under the responsibility given to the National Academy of Sciences by its congressional charter to be an adviser to the federal government and, upon its own initiative, to identify issues of medical care, research, and education. Dr. Harvey V. Fineberg is president of the Institute of Medicine.

The **National Research Council** was organized by the National Academy of Sciences in 1916 to associate the broad community of science and technology with the Academy's purposes of furthering knowledge and advising the federal government. Functioning in accordance with general policies determined by the Academy, the Council has become the principal operating agency of both the National Academy of Sciences and the National Academy of Engineering in providing services to the government, the public, and the scientific and engineering communities. The Council is administered jointly by both Academies and the Institute of Medicine. Dr. Ralph J. Cicerone and Dr. C. D. Mote, Jr., are chair and vice chair, respectively, of the National Research Council.

www.national-academies.org

COMMITTEE ON CAPABILITY SURPRISE ON U.S. NAVAL FORCES

JERRY A. KRILL, Applied Physics Laboratory, Johns Hopkins University, *Co-chair*
J. PAUL REASON, ADM, USN (Retired), Washington, D.C., *Co-chair*
ANN N. CAMPBELL, Sandia National Laboratories
TIMOTHY P. COFFEY, McLean, Virginia
STIRLING A. COLGATE, Los Alamos, New Mexico
CHARLES R. CUSHING, C.R. Cushing & Co., Inc.
SUSAN HACKWOOD, California Council on Science and Technology
LEE M. HAMMARSTROM, Applied Research Laboratory, Pennsylvania State University
NATHANIEL S. HEINER, Northrop Grumman Corporation
LEON A. JOHNSON, Brig Gen, USAFR (Retired), Irving, Texas
CATHERINE M. KELLEHER, University of Maryland and Brown University
JEFFREY E. KLINE, Naval Postgraduate School
ANNETTE J. KRYGIEL, Great Falls, Virginia
THOMAS V. McNAMARA, Raytheon Integrated Defense Systems
RICHARD W. MIES, ADM, USN (Retired), Fairfax Station, Virginia
C. KUMAR N. PATEL, Pranalytica, Inc.
HEIDI C. PERRY, Charles Stark Draper Laboratory, Inc.
GENE H. PORTER, Institute for Defense Analyses
DANA R. POTTS, Lockheed Martin Aeronautics Company
JOHN E. RHODES, LtGen, USMC (Retired), Balboa, California
ROBERT M. STEIN, Brookline, Massachusetts
VINCENT VITTO, Lexington, Massachusetts
DAVID A. WHELAN, The Boeing Company
PETER G. WILHELM, Naval Research Laboratory
JOHN D. WILKINSON, Lincoln Laboratory, Massachusetts Institute of Technology

Staff

CHARLES F. DRAPER, Director, Naval Studies Board
DOUGLAS C. FRIEDMAN, Study Director and Program Officer, Board on Chemical Sciences and Technology (as of October 1, 2012)
BILLY M. WILLIAMS, Study Director (through June 20, 2012)
RAYMOND S. WIDMAYER, Senior Program Officer
MARTA V. HERNANDEZ, Associate Program Officer
SUSAN G. CAMPBELL, Administrative Coordinator
MARY G. GORDON, Information Officer

NAVAL STUDIES BOARD

Marine Corps Liaison

LtGen RICHARD P. MILLS, USMC, Commanding General, Marine Corps Combat Development Command (through August 7, 2013)
LtGen KENNETH J. GLUECK, JR., USMC, Commanding General, Marine Corps Combat Development Command (as of August 8, 2013)

Staff

CHARLES F. DRAPER, Director
RAYMOND S. WIDMAYER, Senior Program Officer
MARTA V. HERNANDEZ, Associate Program Officer
SUSAN G. CAMPBELL, Administrative Coordinator
MARY G. GORDON, Information Officer

Preface

A letter dated December 21, 2011, to National Academy of Sciences President Ralph Cicerone from the Chief of Naval Operations, ADM Jonathan W. Greenert, USN, requested that the National Research Council's (NRC's) Naval Studies Board (NSB) conduct a study to examine the issues surrounding "capability surprise," both operational and technical, facing the U.S. naval services. Accordingly, in February 2012, the NRC, under the auspices of its NSB, established the Committee on Capability Surprise on U.S. Naval Forces.

This committee has found that addressing surprise as it might impact U.S. naval forces is a complex subject with multiple dimensions, including time, mission and cross-mission domains, anticipation of enabling technologies, physical phenomena, and new tactics that can enable surprise. Surprises may come over timescales ranging from seconds to minutes in a complex engagement; alternatively, time may be seen as a cause of evolving, breakthrough surprise that has been secretly developed over decades. Missions such as air defense and undersea warfare, which U.S. naval forces conduct in the open ocean and the littoral regions, all have myriad entry points from which capability surprises can originate (land, air, space, and cyberspace). There are also accelerating new technological advancements globally, which again, singly or in combination, can constitute the basis of a capability surprise.

Given the complexity of surprise, there is no simple way to guard against it. A number of explicit actions are needed. First and foremost, leadership must help others recognize the importance of understanding capability surprise and what it demands of U.S. naval forces, such as ensuring that organizations include preparation for and mitigation of surprise as one of their functions, including scanning and related activities, in order to discern potential surprises. Here, it is important that organizations are timely and diligent in examining the scope and seriousness of such surprises, and that they can identify other organizations that might be able to help anticipate, mitigate, or respond to these surprises.

TERMS OF REFERENCE

At the request of the Chief of Naval Operations, the Naval Studies Board of the National Research Council will conduct a study to examine capability surprise—operationally and technically related—facing U.S. naval forces, i.e., the U.S. Navy, Marine Corps, and Coast Guard. Specifically, the study will

(1) Select a few potential capability surprises across the continuum from disruptive technologies, to intelligence-inferred capability developments, through operational deployments and assess what U.S. Naval Forces are doing (and could do) about these surprises while mindful of future budgetary declines;

(2) Review and assess the adequacy of current U.S. Naval Forces' policies, strategies, and operational and technical approaches for addressing these and other surprises; and

(3) Recommend any changes, including budgetary and organizational changes, as well as identify any barriers and/or leadership issues that must be addressed for responding to or anticipating such surprises including developing some of our own surprises to mitigate against unanticipated surprises.

This 15-month study will produce two reports: (1) a letter report following the third full committee meeting that provides initial observations and insights to each of the three tasks above; and (2) a comprehensive (final) report that addresses the tasks in greater depth.

THE COMMITTEE'S APPROACH

In accomplishing its task, the committee took on a variety of capability surprise topics, as requested in the terms of reference. Today's U.S. naval forces continue to face a wide range of potential threats in the indefinite future and for this reason must continue to balance and meet their force structure needs. Indeed, the *Naval Operations Concept 2010* report—authored by the Chief of Naval Operations, the Commandant of the Marine Corps, and the Commandant of the Coast Guard—noted, among other things, that

> the Naval Service is rebalancing its force structure to address the blue, green and brown water threats potentially posed by very capable state adversaries, as

> well as the maritime security and irregular littoral challenges posed by both state and non-state adversaries.[1]

Included in these envisaged threats are surprises from adversaries employing all sorts of capabilities, from low end to high.

The current study leverages many of the insights from the 2009 Defense Science Board (DSB) report and the 2008 Naval Research Advisory Committee (NRAC) report but focuses on U.S. naval forces. It is divided into two parts. The first part selects a few surprises from across a continuum of surprises, from disruptive technologies, to intelligence-inferred capability developments, to operational deployments, and assesses what the Navy, Marine Corps, and Coast Guard are doing (and could do) about them while being mindful of future budgetary declines. The second part examines which processes are in place or could be in place in the Navy, the Marine Corps, and the Coast Guard to address such surprises. For example, it explores the pros and cons of a variety of ways to improve the naval forces' response to such surprises by means of red teaming, by employing our own capabilities to surprise others, and by identifying barriers that could prevent the adoption of such processes or reduce their effectiveness.

The committee was convened in February 2012. After its first three meetings, the committee drafted its interim report.[2] It held three additional meetings and conducted a site visit over the next 4 months to gather input from the relevant communities and to discuss its findings and recommendations. An outline of the committee's meetings is provided in Appendix F.

ORGANIZATION OF THE REPORT

This final report contains the committee's findings and recommendations and builds on the framework first described in the interim report. Chapter 1 provides background information and introduces the six phases that are proposed to address capability surprise. Its Finding 1 and Recommendation 1 are complemented and supported by findings and recommendations that are found in the subsequent chapters. Each of the next six chapters—Chapters 2 through 7—describes the stakeholders, performers, and activities of six functional framework phases. Those chapters also describe the importance and key attributes of these six phases in the context of the surprise scenarios and exemplars, which are described in more detail in Chapter 8 and Appendixes A and B.

Finally, in Chapter 8, summarized in "ready reference" format, the committee presents the composite capability described in the preceding chapters in terms

[1]Gen James T. Conway, USMC; ADM Gary Roughead, USN; and ADM Thad W. Allen, USCG. 2010. *Naval Operations Concept 2010*, Department of the Navy, Washington, D.C., p. 82.

[2]The interim report, *Capability Surprise for U.S. Naval Forces: Initial Observations and Insights,* was released on January 15, 2013.

of modifications that leverage the entire naval infrastructure to address surprise on a routine basis with adequate, prioritized resources. The result is expected to be a change in naval culture to ensure more responsive, more resilient, and more adaptive behavior across the organization from the most senior leadership to the individual sailors, Marines, and Coast Guardsmen.

Jerry A. Krill, *Co-chair*
J. Paul Reason, *Co-chair*
Committee on Capability Surprise on U.S. Naval Forces

Acknowledgment of Reviewers

This report has been reviewed in draft form by individuals chosen for their diverse perspectives and technical expertise, in accordance with procedures approved by the National Research Council's Report Review Committee. The purpose of this independent review is to provide candid and critical comments that will assist the institution in making its published report as sound as possible and to ensure that the report meets institutional standards for objectivity, evidence, and responsiveness to the study charge. The review comments and draft manuscript remain confidential to protect the integrity of the deliberative process. We wish to thank the following individuals for their review of this report:

Ruth A. David, ANSER (Analytic Services, Inc.);
David J. (Jack) Dorsett, VADM, USN (Retired), Northrop Grumman Corporation;
Robert A. Frosch, Harvard University;
Wayne P. Hughes, Jr., Naval Postgraduate School;
Harry W. Jenkins, Jr., MajGen, USMC (Retired), Gainesville, Virginia;
Kathryn B. Laskey, George Mason University;
Cato T. Laurencin, University of Connecticut;
D. Brian Peterman, VADM, USCG (Retired), Command at Sea International;
Fred E. Saalfeld, Springfield, Virginia;
Nils R. Sandell, Jr., Concord, Massachusetts; and
Neil G. Siegel, Northrop Grumman Information Systems.

Although the reviewers listed above provided many constructive comments and suggestions, they were not asked to endorse the conclusions or recommendations, nor did they see the final draft of the report before its release. The review of this report was overseen by Stephen M. Robinson, University of Wisconsin-Madison. Appointed by the National Research Council, he was responsible for making certain that an independent examination of this report was carried out in accordance with institutional procedures and that all review comments were carefully considered. Responsibility for the final content of this report rests entirely with the authoring committee and the institution.

Contents

APPENDIXES

Summary

In response to the Chief of Naval Operations (CNO), the National Research Council appointed a committee operating under the auspices of the Naval Studies Board to study the issues surrounding capability surprise, both operational and technical, facing U.S. naval forces—the Navy, the Marine Corps, and the Coast Guard.[1] Capability surprise is both inevitable and inherently complex. It has multiple dimensions, including time, mission, and cross-mission domains; anticipation of enabling technologies; physical phenomena; and new tactics that may enable surprise.[2] Anticipating and mitigating capability surprise may seem daunting for U.S. naval forces in today's evolving national security environment. However, many efforts are already under way that could be leveraged to bring about an increase in preparedness against surprise through focus, culture, and U.S. naval leadership.

DEFINING "SURPRISE" AND STUDY SCENARIOS

From a military operational context and for the purposes of this report, surprise is an event or capability that could affect the outcome of a mission or

[1]Throughout this report, the terms "Navy," "Marine Corps," and "Coast Guard" are used. Unless stated otherwise, these refer to the U.S. Navy, the U.S. Marine Corps, and the U.S. Coast Guard.

[2]Surprises may come over timescales ranging from seconds to minutes in a complex engagement, to the evolving, breakthrough surprise that might have been secretly developed over decades. The mission domains such as air defense and undersea warfare, which U.S. naval forces operate across the open ocean and littoral (land, air, space, and cyberspace), all have myriad entry points from which capability surprises can originate. There are also accelerating new technological advancements globally, which singularly or in combination can constitute the basis of a capability surprise.

campaign for which preparations are not in place. By definition, it is not possible to anticipate true surprise. It is only possible to minimize the number of possible surprises by appropriate planning, to create systems that are resilient to the unexpected actions of an adversary, and to rapidly and effectively respond when surprised. There are two classes of surprise that fall within this military operational context and conform to the terminology provided in the study's terms of reference:[3] (1) intelligence-inferred surprise and (2) disruptive technology and tactical surprise.

"Intelligence-inferred surprise" is an event or capability developed over a relatively long time line—years—that U.S. naval forces were aware of in advance of its looming operational introduction, but for which they might not have adequately prepared. "Disruptive technology" (including the disruptive application of existing technology) and "tactical surprise" are short-time-line—hours to months—events or capabilities for which naval forces probably have not had sufficient time to develop countermeasures. In some cases both types of surprise can occur—for example, a much anticipated surprise capability that is found on the battlefield to have tactical war reserve modes.

Two variants of disruptive technology and tactical surprise have been identified. The first is the pop-up of a new capability enabled by a new technology or an unexpected application of an existing well-known technology—for example, improvised explosive devices (IEDs) as well as unexpected tactics. The second variant, black swan events—may be "self-inflicted" surprises—for example, blind spots or vulnerabilities in our own systems. Here, no amount of surveillance would have allowed anyone to predict the event.[4] Such events may also be the result of a sudden U.S. policy change or directed action, such as Operation Burnt Frost,[5] or natural disasters that are to be anticipated but not on such an extreme scale—for example, the March 2011 Fukushima Disaster.[6]

Sometimes the mitigation of surprise will take the form of naval forces

---

[3]The study's terms of reference (TOR) are provided in the Preface. The TOR charged the committee to produce two reports over a 15-month period. The present report, the committee's final report, accords with the committee's interim report and contains similar text; however, it provides specific findings and recommendations along with additional analysis.

[4]Nassim Taleb defines a black swan as "a highly improbable event with three principal characteristics: It is unpredictable; it carries a massive impact; and, after the fact, we concoct an explanation that makes it appear less random, and more predictable, than it was." For additional reading on black swan events, see Nassim Nicholas Taleb, 2010, *The Black Swan: The Impact of the Highly Improbable*, 2nd edition, Random House Trade Paperbacks, New York, N.Y.

[5]Operation Burnt Frost was the mission to shoot down a nonfunctioning National Reconnaissance Office Satellite in 2008. RADM Brad Hicks, USN, Program Director, Aegis Ballistic Missile Defense, "Aegis Ballistic Missile Defense: Press Briefing, March 19, 2008," presented to the committee by RADM Joseph A. Horn, Jr., USN, Program Executive, Aegis Ballistic Missile Defense, and Conrad J. Grant, Johns Hopkins University Applied Physics Laboratory, May 16, 2012, Washington, D.C.

[6]A partial profile of U.S. naval response to the Fukushima disaster—a combined earthquake, tsunami, and nuclear reactor catastrophe—in a coordinated effort known as Operation Tomodachi is found at http://www.nbr.org/research/activity.aspx?id=121. Accessed June 13, 2012.

developing their own surprises as potential counters. Throughout this report, when the committee refers to "mitigation," the term includes not only measures to counter the potential surprises of an adversary but also our own surprises delivered preemptively.

As discussed at length throughout the report, the committee recognizes that surprise cannot be completely anticipated. One cannot, for example, anticipate precisely how even an intelligence-inferred surprise will unfold. Therefore the resilience of U.S. naval forces and their capabilities will remain key. Resilience takes a number of forms:

1. Design features in mission systems to counter or protect against such surprise vectors as electronic countermeasures and cyberattacks.
2. War reserves in the form of backup systems or frequency bands in case a primary system or channel is rendered inoperative.
3. Exercises to explore alternative tactics, techniques, and procedures for use if and when adversarial surprises are unleashed.
4. Training for warfighters and mission operators that incorporates surprise elements not only to develop resourcefulness in surprise mitigation but also to instill the confidence that surprises usually have work-arounds.
5. As part of our proficiency training, creation of our own countersurprises to give pause to an adversary's tactics and buy time to counter its surprises.
6. Provision of design margins (such as power and space) on platforms to support rapid fielding capabilities as necessary.
7. Plan for low-cost countersurprise tactics that use nonkinetic effects such as deceptive electronic countermeasures, decoys, and cyber operations.
8. Development of contingency plans for deployment of special operations forces.

In accordance with the terms of reference and to guide its analysis and, ultimately, identify findings and propose recommendations to U.S. naval leadership, the committee selected three surprise scenarios:

- Scenario 1: Denial of space access;
- Scenario 2: An asymmetric engagement with complex use of cybermethods in a naval context; and
- Scenario 3: A black swan event to which the front-end scanning and prioritization framework for mitigating surprise is not applicable.

In short, these scenarios aided in studying what U.S. naval forces are doing and could do to anticipate and respond to capability surprise and to mitigate it. These scenarios were chosen in part because they address issues important to the U.S. Navy, Marine Corps, and Coast Guard. Denial of space access is treated

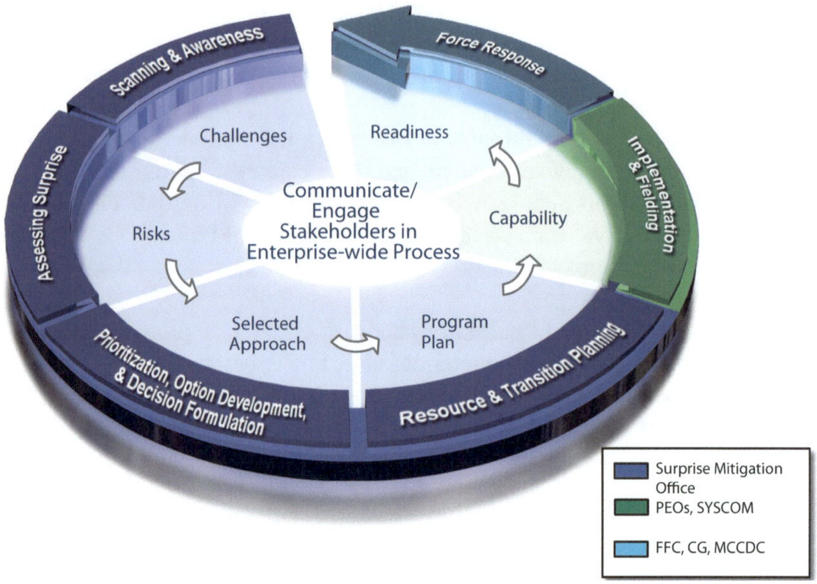

FIGURE S-1 Recommended framework for addressing capability surprise.

in a Navy context, social media manipulation is framed to suit a Marine Corps environment, and disaster relief would typically involve the Coast Guard.

A FRAMEWORK FOR ADDRESSING SURPRISE

With the above-mentioned scenarios in mind as well as exemplars for a military response, which will be introduced later, and to guide U.S. naval forces in addressing a potential capability surprise, the committee adapted a functional framework consisting of six phases that can be aligned with the development functions, accountabilities, and principles required for effective surprise mitigation.

The framework, shown in Figure S-1 and described in greater detail throughout this report, comprises six phases: (1) Scanning and Awareness; (2) Assessing Surprise; (3) Prioritization, Option Development, and Decision Formulation; (4) Resource and Transition Planning; (5) Implementation and Fielding; and (6) Force Response.

Options for Coordinating Surprise Mitigation

In order for leadership to enable U.S. naval forces to implement and manage surprise mitigation, virtually every development and acquisition program conducts some form of the following:

- Threat-response activities, including the projection of future threats,
- Development effort directed at meeting the projected threats, and
- Continual assessment of existing and developmental system designs for potential vulnerabilities so that corrective measures can be undertaken.

At the same time, a number of mission or mission-support capabilities are of primary concern:

- There is presently no authorized office to address these concerns,
- There are areas of vulnerability in programs for which no executive has been identified to coordinate the surprise mitigations,
- The potential impact on mission capabilities of surprises that pop up from emerging technologies or applications of existing technology is at present ambiguous, and
- There is no assurance that sufficient attention is paid to identified risks.

Finding 1: Capability surprise is both inevitable and inherently complex, and it requires U.S. naval forces to engage in a broad spectrum of issues, from horizon scanning to red teaming to experimentation and rapid prototyping, to exercising, fielding, and training. While there are a few exemplar organizations in the Navy, Marine Corps, and Coast Guard as well as the Department of Defense that effectively work on capability surprise, there is neither an overall framework for nor a clear delineation of U.S. naval forces responsibilities.[7]

Recommendation 1: The Chief of Naval Operations (CNO), the Commandant of the Marine Corps (CMC), and the Commandant of the Coast Guard (CCG) should establish a common framework for U.S. naval forces to address capability surprise, as shown in Figure S-1.

The CNO, in concert with the CMC and the CCG, should establish and fund a surprise mitigation office to serve as the executive agent for addressing capability surprise for U.S. naval forces. Specifically, this office should comprise a set of organic operational, technical, assessment, and intelligence-oriented staff and draw input and analysis with a global, multicultural perspective from a multitude of communities: operational; intelligence; acquisition, research, and development; system commands; program executive offices; war colleges; military fellows; government laboratories; Federally Funded Research and Development Centers; university-affiliated research centers; industry; and academia. To ensure that U.S. naval forces are proactive in developing and anticipating surprise capabilities it is recom-

[7]Some exemplars identified by the committee include the Navy's SSBN Security program, the Air Vehicle Survivability Evaluation Program (Air Force Red Team), and the Aegis Ballistic Missile Defense program (in response to Operation Burnt Frost).

mended that representatives from Special Operations Command (SOCOM) be an integral part of this office. Based on the CNO's decision above on which option to pursue (see Table S-1), the office should support the timely transition of surprise capabilities to the appropriate organizations of primary responsibility and monitor the transition to, and effectiveness within, operational forces.

To meet the committee's objective of accountable attention to surprise, especially for nonacquisition programs as outlined above, eight "office options" were considered (Table S-1). Each of these eight options is described in detail in Chapter 1.

After weighing the pros and cons of each of the eight office options, the committee identified the N9I office as the existing organization best equipped to coordinate consideration of surprise for U.S. naval forces and work in concert with joint forces and national assets. The N9I office currently ensures integration across programs, i.e., warfare systems. Although its main role is to oversee integrated financial and budget management, the N9I office also manages the Naval Integrated Fire Control-Counter Air (NIFC-CA) program. This program is particularly relevant because it requires the technical and operational integration of sea surface, land, and airborne combatants (Aegis, Army Aerostat, and E-2D) sharing information over networks—Link 16 and cooperative engagement capability (CEC)—to enable weapons such as the Standard Missile 6 to engage airborne cruise missiles over terrain that might be beyond the horizon of the firing unit, using composite track and targeting data. The ability of N9I to coordinate programs is essential for a capability surprise office dealing with surprises that are probably not addressed by a single new program or capability upgrade. To extend the N9I office's mission to also mitigate capability surprise seems synergistic, but it will need additional staffing and coordination.

TABLE S-1 Office Options Considered for Surprise Mitigation for U.S. Naval Forces

Option No.	Option[a]
1	Incorporate into existing N9I office
2	Incorporate into existing N2/N6I office
3	Establish new center of excellence
4	Assign to Office of Naval Research
5	Create rapid acquisition office with PEO
6	Create new OPNAV office
7	Delegate to OSD surprise office
8	Incorporate into existing DASN (RDT&E) office

[a]OPNAV, Office of the Chief of Naval Operations; PEO, program executive office; OSD, Office of the Secretary of Defense; DASN (RDT&E), Deputy Assistant Secretary of the Navy for Research, Development, Testing and Evaluation.

The committee was especially mindful that in the present era of limited resources and the need to consider affordability, a new office would be unrealistic. On the other hand, the committee was keenly aware it would be ineffective to simply declare the need for the framework and to advise the naval culture that it should embrace the consideration of potential surprises. As pointed out directly by the Defense Science Board, "Rarely is there a case of true surprise. Post mortems almost always identify that someone has provided warning, but that warning was not heeded."[8] Therefore, an office with sufficient authority and connectivity as defined by the recommended framework, is necessary to effectively address surprise.

The committee considered that it would be presumptuous to formally recommend option 1 if CNO determines, based on plans and considerations of which the committee is not aware, that another office is more appropriate. Therefore, the committee recommends the establishment of a surprise mitigation office and considers N9I the most likely organization to lead it for U.S. naval forces. It understands, however, that the CNO may identify a more appropriate entity to take on this role according to the framework presented and recommended in this report.

PRIORITY AREAS FOR ACTION: FINDINGS AND RECOMMENDATIONS 2 THROUGH 6

The following five areas are key to successful surprise mitigation. Each is an integral part of the framework introduced earlier in this summary and discussed in detail in the report.[9]

Expand and Create Roles for Greater Scanning and Awareness

Finding 2: The Office of Naval Research-Global (ONR-G) is focused on scanning at the 6.1 and 6.2 levels; the Office of Naval Intelligence (ONI) on technical intelligence, primarily systems in development, testing, or operational exercises (6.3-6.7 levels); and fleet intelligence on observed operational behavior. However, there is no integrated, comprehensive scanning that also explores the linking of these observations and the emergence of consumer technologies of potential impact. While there are usually early indicators of potential capability surprises, there does not appear to be a coordinated means for U.S. naval forces to explicitly scan the horizon for such indicators; to capture, retain, and vet such indicators with relevant organiza-

[8]Defense Science Board. 2009. *Report of the Defense Science Board 2008 Summer Study on Capability Surprise*, Volume I: Main Report, Office of the Under Secretary of Defense for Acquisition, Technology and Logistics, Washington, D.C., September, p. xiii.

[9]The findings and recommendations presented here are expanded on in Chapter 8 of this report, "Putting It All Together." That chapter summarizes the essential elements of Chapters 2 through 7 and provides a context for the findings and recommendations presented here.

tions; and, ultimately, to inform senior leadership of potential capability surprises. Such a coordinated means would need to scan, recognize, categorize, analyze, and report technical and/or operational surprises on a global basis.

Recommendation 2: Using existing fleet resources, the Chief of Naval Operations should enlist the support of the combatant commanders and their naval component commands in order to scan, recognize, capture, and report potential capability surprises outside the continental United States (OCONUS). Most notably, the Commander, U.S. Naval Forces Europe and Africa, and the Commander, U.S. Pacific Fleet, should establish comparable counterparts within their respective staffs to the surprise mitigation office referred to in Recommendation 1.

To further aid in identifying potential capability surprises—technological, operational, and/or otherwise—the Assistant Secretary of the Navy for Research, Development and Acquisition (ASN RDA) should (1) appoint the Deputy Assistant Secretary of the Navy for Research, Development, Testing and Evaluation (DASN RDT&E) as the Navy Department's Chief Technology Officer, with primary responsibility for providing technical advice for all phases of the framework for the surprise mitigation office in Recommendation 1 and (2) direct the Chief of Naval Research to establish a "virtual" scanning and awareness structure led by the Office of Naval Research-Global, engaging the technical, intelligence, and operational communities in order to systematically scan the horizon, maintain awareness, and conduct technology readiness assessments for both the CNO and the surprise mitigation office, as called for in Recommendation 1.

Improve Methodologies for Assessing Surprise

Finding 3: Organizations that anticipate and respond effectively to potential capability surprises—such as the Navy's SSBN[10] Security program, the Air Vehicle Survivability Evaluation program (Air Force red team), and the Aegis Ballistic Missile Defense program (in response to Operation Burnt Frost)—appear to possess the following characteristics: senior leadership support; team independence; access to a strong base of cross-disciplinary technical and operational expertise; an ability to identify threats through campaign-level modeling, system-of-systems simulation, and high-fidelity physics-based models; precise vulnerability modeling and analysis capabilities validated by test and experiment data; mechanisms to recommend and/or deploy solutions as necessary; adequate, steady funding; and focus on a particular mission such as Navy SSBN Security. In addition, these organizations appear to leverage modeling, simulation, and analysis tools in conjunc-

[10]SSBN, nuclear-powered ballistic missile submarine.

tion with a network of experts to expose bias, offer critical review, model and test against potential vulnerabilities, and demonstrate alternative solutions to respond to surprise.

At the same time, assessments of threats to the critical technologies that enable U.S. naval forces appear to be conducted on a small scale rather than being quantified, modeled, and characterized for U.S. naval forces as a whole. For example, threats to precision navigation and timing sources or cyberattacks embedded within Navy weapon systems could impact a wide array of naval operations. However, U.S. naval forces as a whole do not seem to be utilizing the best methodologies for assessing surprise. One of these methodologies would be the creation of red teams, that could simulate or represent adversarial thinking across global cultures.

Recommendation 3: As its first tasking from the Chief of Naval Operations (CNO), the surprise mitigation office (see Recommendation 1) should (1) identify and prioritize any broad response to operational and technology threats that are not owned by any one mission authority and (2) establish threat study groups to characterize, quantify, and model these specific threats as well as leverage existing resources (modeling, simulation, and analysis tools and test data used by a network of subject matter experts in academia, industry, laboratories, and the Service colleges). The output from (1) and (2) should be disseminated to U.S. naval leadership as soon as possible. Careful attention should be paid to surprises not addressed by any program, or where a substantial gap exists between programs.

The CNO, the Commandant of the Marine Corps (CMC), and the Commandant of the Coast Guard (CCG) should take steps to ensure that red teams—with sufficient independence and expertise but, at the same time, and a fresh influx of participants—are able to depart radically from traditional thinking in order to help U.S. naval forces as a whole prepare for combat and develop new tactics. In particular, efforts should be made to expand and periodically refresh the composition of red teams to achieve a greater diversity in thinking and better represent the adversary.

Work Joint and Naval Solutions for Responding to Surprise

Finding 4: With the recent establishment of the Strategic Capabilities Office within the Office of the Under Secretary of Defense for Acquisition, Technology, and Logistics (USD AT&L), there appears to be an opportunity for a surprise mitigation office to provide naval force component solutions to surprises facing the entire Joint Force. Furthermore, there is an opportunity for both offices to draw on each other by sharing expertise, methodologies (modeling, simulation, analysis, red teaming), and learning.

Recommendation 4: The Chief of Naval Operations (CNO) and the Under Secretary of Defense for Acquisition, Technology, and Logistics (USD AT&L) should encourage their respective "surprise offices" to develop and foster a close working relationship with each other. In particular, the CNO and USD AT&L should direct their surprise offices to share ideas and to collaborate on methodologies (modeling, simulation, analysis, test data, and red teaming) in the interest of efficiency and obtaining consistent and coordinated results. Technical interchange meetings and the frequent exchange of information between the two offices and others that may be eventually established by the other Services should also be encouraged.

Plan for Surprise

Finding 5: In planning for surprise, deficits appear to have arisen in two areas:

a. Management processes by which resourcing decisions are made and that potentially impact preparations for capability surprise among U.S. naval forces appear to be inadequate and to lack reserve capacity. In particular, there appears to be limited flexibility in the way of design margins for platforms and payloads to respond to a range of potential capability surprises, and it further appears that the Department of the Navy's investment in science and technology is insufficient to provide a robust array of technology building blocks that allow a rapid response to a broad range of potential surprises.

b. In addition, the Department of the Navy is not extending the full measure of open architecture principles throughout system development and deployment life cycles nor is it making best use of permissible contracting exceptions or best acquisition practices in responding to potential capability surprise in a timely and efficient manner.

Recommendation 5: In planning for surprise,

a. The Chief of Naval Operations (CNO), the Commandant of the Marine Corps (CMC), and the Commandant of the Coast Guard (CCG) should add to their respective program planning guidance an explicit provision(s) that allows them to resource unit considerations and equipment design specifications, such as adding some adequate design margins into platforms and payloads in response to potential capability surprise.

b. The Chief of Naval Research (CNR) should invest in discovery and invention (6.1 and early 6.2) research areas that take advantage of the entire payload value chain (i.e., payloads versus platforms; modularity versus integration; and reprogrammability), and inclusion of appropriate software and hardware design margins into development requirements. The Assistant

Secretary of the Navy for Research, Development, and Acquisition (ASN RDA) should ensure that acquisition and contracting personnel are trained in the development of threshold versus objective requirements, the unique requirements associated with the use of commercial products, and the appropriate use of the waiver process in tailoring responses to potential capability surprise.

c. The surprise mitigation office (see Recommendation 1) should encourage broader cross-organizational pre-planning in anticipation of, and based on previous, black swan events that can cut across U.S. government department responsibilities, and it should also serve as the lead resource officer for the rapid fielding of new capabilities to counter unanticipated surprises.

Prepare for Surprise Now

Finding 6: In preparing for surprise, it appears that

a. U.S. naval forces are not preparing adequately for potential capability surprise in current exercises and experiments. For example, naval exercises do not usually accommodate degraded environments or unexpected developments to be realistically addressed, and training tends to focus on current operations and leaves inadequate time for experimentation and use of new technologies.

b. U.S. naval forces do not have an advocate and resource sponsor to rapidly field new capabilities to counter pop-up surprises, nor are they taking advantage of any existing capabilities, as identified in the Navy Readiness Reporting Enterprise (NRRE), that could potentially counter surprises of all types.

c. It is unclear whether some key classified capabilities—to the extent any exist—are disclosed to planners or operators and therefore may not be routinely incorporated into combatant plans or practiced by operators.

Recommendation 6: In preparing for surprise,

a. Operational commanders should incorporate, when feasible, degraded environments and aspects of surprise into exercise and training scenarios to improve preparation for and response to surprise. U.S. Fleet Forces Command (FFC), including directed type commanders, Marine Corps Combat Development Command (MCCDC), and U.S. Coast Guard Force Readiness Command (FORCECOM), should expand experimentation and related activities to create concepts and tactics to counter surprise. To offset resource impacts, activities of limited scope, such as small-unit or small-scale experiments, may be utilized.

These commanders should use the results from exercises and experimentation to analyze and assess their preparedness for capability surprise. As appropriate, they should formulate measures and incorporate them into the existing readiness reporting structures through the appropriate naval organizations and readiness reporting systems.

The results of incorporating surprise into exercises and experimentation will be forwarded to the appropriate Service organizations, including the capability surprise office, and entered into the training continuum, as appropriate.

b. Commander, U.S. Fleet Forces Command (FFC), should leverage the Navy Readiness Reporting Enterprise (NRRE) and provide operational commanders with any existing capabilities that could counter surprises of all types.

c. Finally, operational commanders should work to ensure that any of our key classified capabilities—to the extent any exist—are disclosed to planners or operators so that they are incorporated into combatant plans or practiced by operators in responding to capability surprise.

ROLES AND ACTIVITIES TO ADDRESS CAPABILITY SURPRISE

Figure S-2 summarizes the outputs; owners, stakeholders, and participants/performers; and activities in each phase of the framework. It is a compilation of the narratives in the chapters that follow. The stakeholders and participants are brought together, as appropriate, to assess signs of emerging surprises detected by the scanning and awareness activities of the operational, research, and intelligence establishments in Phase 1 and to move toward the modeling and assessment organizations and laboratories to verify feasibility in the middle phases. Finally, the acquisition and operational organizations are the primary players in the final phases. The activities of each phase are summarized, and the primary output of each phase is shown in the first row of the table, from identification of challenges to delivery of capability and ensuring of deployed readiness. Sometimes with disruptive surprises recognized only in the heat of an operation, a crash program will be initiated that involves only phases 4, 5, and 6, because the early observations have been preempted by new findings on the battlefield or disaster area.

The committee has determined that many or most of the requisite functions already reside within the U.S. Navy, Marine Corps, and Coast Guard. However, they are not sufficiently integrated, prioritized, or advocated. This determination is the primary motivation for Recommendation 1—namely, that a surprise mitigation office be established to coordinate and prioritize surprise mitigation for U.S. naval forces. The office would serve as the executive agent for the first four phases. The final two phases of the framework would be ably led by the identified existing organizations once the surprise mitigation office has prioritized, defined, and planned appropriate measures for its participation in these final phases to

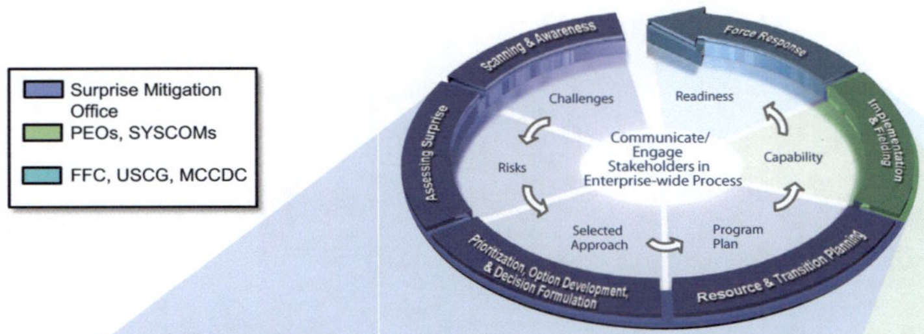

	Scanning & Awareness	Assessing Surprise	Prioritization, Option Development & Decision Formulation	Resource & Transition Planning	Implementation & Planning	Force Response
Outputs	Challenges	Risk	Priority, feasibility, & affordability	Program plan	Capability	Readiness
Owner(s)	Surprise Mitigation Office	Surprise Mitigation Office	Surprise Mitigation Office	Surprise Mitigation Office	PEOs SYSCOMs	FFC USCG FORCECOM MCCDC
Stakeholders	ONR-G ONI FFC USCG FC-A MCCDC	N81 FFC USCG FC-A MCCDC	N2/N6 N8 N9 DASN RDTE/CTO ONR/CNR MCCDC USCG FC-P	DASN RDTE/CTO MCCDC USCG CG-8/ FORCECOM SYSCOMs	Surprise Mitigation Office N9 N2/6 DASN RDTE/CTO MCCDC USCG FC-P SYSCOMs	OSD/SCO Surprise Mitigation Office USCG FC-E & FC-T
Participants/ Performers	Industry Labs/Warfare Centers Academia IC	NWDC, MCWL Industry Labs/Warfare Centers Think tanks Services' Colleges	NWDC, MCWL Industry Labs/Warfare Centers Think tanks Services' Colleges	Industry Labs/Warfare Centers PEOs SYSCOMs ONR	OPTEVFOR Industry Labs/Warfare Centers	USN, USMC, & USCG Operational Forces
Activities	Technical Intelligence S&T monitoring Physical modeling	Campaign models Cultural assessments System modeling	TTP Modeling Prototype Experiments/ATD	Program Planning, Budgeting, & Coordination	TTP/CONOPS System Modifications New Programs	TTP/CONOPS Assessment Testing Training Exercises Equipping

FIGURE S-2 Roles and activities to address capability surprise. SYSCOM, Systems Command; IC, intelligence community; NWDC, Navy Warfare Development Command. See Appendix D for the definitions of acronyms that have not yet been spelled out.

ensure that the surprise mitigation capabilities are deployed in a suitable and timely manner.

Figure S-3 maps the recommendations to organizations within and outside the U.S. Navy, Marine Corps, and Coast Guard. As shown from the mapping, the impact of each recommended change on any part of the organization is rather small; however, in the committee's view, the sum of each of these changes would result in a much more integrated and prioritized approach to addressing capability surprise for U.S. naval forces.

In summary, the framework and organizational recommendations will enhance the ability of U.S. naval forces to prepare for capability surprise. The recommendations will further support the U.S. naval leadership and enhance the Navy's culture of awareness and its proactive focus on becoming more resilient against surprise.

The report concludes with a ready reference describing how the recommendations would be implemented across the enterprise based on the details in the chapters.

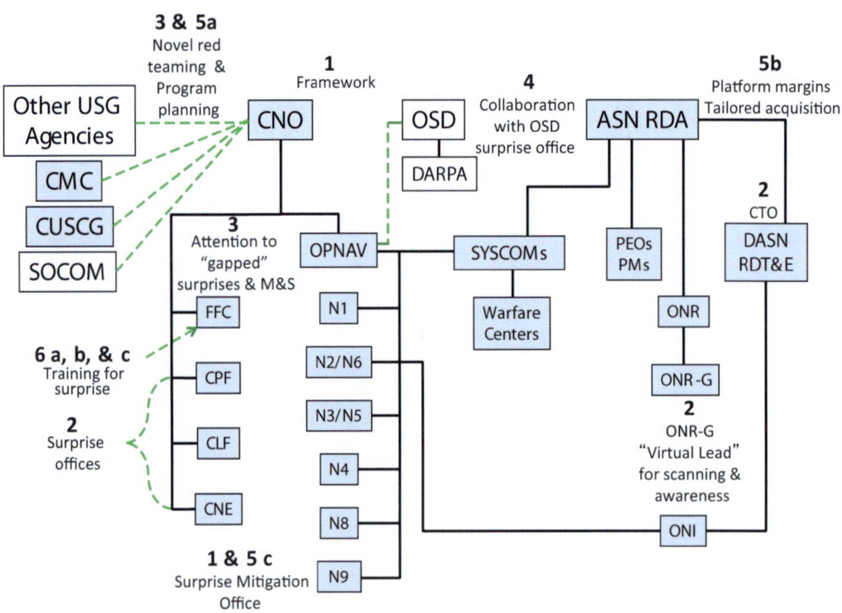

FIGURE S-3 Mapping of recommendations to organizations within and outside the U.S. Navy, Marine Corps, and Coast Guard. Many of the acronyms have already been identified in the text. See Appendix D for the definitions of those that have not so far been spelled out.

1

Framing the Problem

BACKGROUND

Recent reports have addressed the issue of surprise, though not specifically as it relates to U.S. naval forces (the Navy, the Marine Corps, and the Coast Guard). The 2009 Defense Science Board (DSB) Summer Study on Capability Surprise noted, among other things, as follows:

> Surprise can spring from many sources. It can arise in the laboratory—a result of scientific breakthrough. It can arise during the transition from concept to fielded product: rapid fielding of the same technology can create tremendous advantage to whoever fields the system first. It can also arise when an existing capability is employed in an unconventional way or when low-end technology is adapted in unforeseen ways that create an effective capability against high-end U.S. systems.[1]

In short, the DSB report reviewed many historical surprises to our nation and categorized them as either "known surprises" (i.e., surprises that should have been anticipated and acted upon because they were clearly in the offing) or "surprising surprises" (i.e., those that may have been anticipated by some but were not addressed—swamped as these persons or institutions were by thousands of other surprise possibilities—or those that actually were true surprises).[2]

---

[1]Defense Science Board. 2009. *Report of the Defense Science Board 2008 Summer Study on Capability Surprise*, Volume I: Main Report, Office of the Under Secretary of Defense for Acquisition, Technology and Logistics, Washington, D.C., September, pp. vii and viii.

[2]The temporal and impact aspects of capability surprises vary widely and call for different approaches to prepare for and respond to them. As additional background for this study, the committee examined several historical examples of "surprises" that have had significant impact on naval and

In addition, a 2008 Naval Research Advisory Committee (NRAC) report entitled *Disruptive Commercial Technologies* noted, among other things, that "the internet functions effectively as both an R&D resource and supply chain for irregular forces throughout the world. Commercial technologies pose a real and enduring threat to Marine forces."[3] The NRAC report concluded that there exist globally available commercial technologies that could be used in hostile ways against Marine forces. While the NRAC report did not focus on technology surprise per se, it did examine the power of unconventional and unconstrained imagination that can be used against Marine forces operating around the world.

DEFINING "SURPRISE"

From a military operational standpoint and for the purposes of this report, surprise is an event or capability that could affect the outcome of a mission or campaign for which preparations are not in place. By definition, it is not possible to truly anticipate surprise. It is only possible to prevent it (in the sense of minimizing the number of possible surprises by appropriate planning), to create systems that are resilient to an adversary's unexpected actions, or to rapidly and effectively respond when surprised. The committee identified two classes of surprise that fall within this military operational context and described them using the terminology in the study's terms of reference and in the interim report: (1) intelligence-inferred surprise and (2) disruptive technology and tactical surprise.

Intelligence-inferred surprise is an event or capability that developed over a relatively long time—years—and that naval forces were aware of in advance of its operational introduction but for which they may not have adequately prepared. Disruptive technology (including the disruptive application of existing technology) and tactical surprise involve short-timeline events or capabilities—hours to months—for which naval forces probably have not had sufficient time to develop countermeasures unless they were at least somewhat anticipated. A surprise may fit in both categories—for example, Blitzkrieg warfare combined the latest tank and aircraft technology with surprise penetration tactics.

Much intelligence-inferred surprise is being continually monitored by naval program areas such as air and missile defense, antisubmarine warfare, and strike warfare systems. Here, the future threat is projected, and upgrades to naval systems to meet the threat are being developed and fielded. This report will not so

military operations, including short-lived surprises (such as the suicide bomb attacks on the USS *Cole* and the 9/11 World Trade Center) and longer-term surprises, resulting in major changes in U.S. naval and military forces (such as the battle between the *Monitor* and *Merrimac,* the first-ever battle between ironclad warships), as well as Russia's launch of Sputnik (a surprising use of space, leading to the creation of the Defense Advanced Research Projects Agency [DARPA]) and Germany's Blitzkrieg, uniquely combining and exploiting the capabilities of known entities.

[3]Naval Research Advisory Committee. 2008. *Disruptive Commercial Technologies*, Office of the Assistant Secretary of the Navy for Research, Development and Acquisition, Washington, D.C., June 26, p. 15.

much deal with intelligence-inferred surprises, which are already being addressed (except those that were selected as exemplars of key programs enabling naval capabilities to keep abreast of threat and technology opportunities). Instead, it deals with intelligence-inferred surprises for which cradle-to-grave upgrades do not exist or whose defeat requires coordination among a number of naval programs. One example of such an orphan intelligence-inferred surprise—denial of space access—is brought up throughout this report.

Two variants of disruptive technology and tactical surprise have been identified. The first is the pop-up of a new capability enabled by a new technology or an unexpected application of an existing well-known technology—for example, new kinds of triggers for improvised explosive devices (IEDs) as well as unexpected tactics. The second variant, black swan events, may be "self-inflicted" surprises—for example, blind spots or vulnerabilities in our own systems. Here, no amount of surveillance would have allowed anyone to predict the event.[4] Such events may be the result of a sudden U.S. policy change or directed action, such as Operation Burnt Frost,[5] or natural disasters that are to be anticipated but not on such an extreme scale—for example, the March 2011 Fukushima Disaster.[6]

In the broadest sense, surprise grants an adversary the chance to take unexpected action or to produce consequences that we did not prepare for. Surprises may also result from operational, social, natural, or political factors for which technology or lack of mitigating technology may not be the primary impactor.

INITIAL OBSERVATIONS

The Committee on Capability Surprise on U.S. Naval Forces has found that addressing surprise as it might impact U.S. naval forces is a complex subject with multiple dimensions, including time, mission and cross-mission domains, anticipation of enabling technologies, physical phenomena, and new tactics that can enable surprise. Surprises may come over timescales ranging from seconds up to minutes in a complex engagement; alternatively, time may be seen as a cause of evolving, breakthrough surprise that has been secretly developed over decades.

---

[4]Nassim Taleb defines a black swan as "a highly improbable event with three principal characteristics: It is unpredictable; it carries a massive impact; and, after the fact, we concoct an explanation that makes it appear less random, and more predictable, than it was." For additional reading on black swan events, see Nassim Nicholas Taleb, 2010, *The Black Swan: The Impact of the Highly Improbable*, 2nd edition, Random House Trade Paperbacks, New York, N.Y.

[5]Operation Burnt Frost was the mission to shoot down a nonfunctioning National Reconnaissance Office Satellite in 2008. RADM Brad Hicks, USN, Program Director, Aegis Ballistic Missile Defense, "Aegis Ballistic Missile Defense: Press Briefing, March 19, 2008," presented to the committee by RADM Joseph A. Horn, Jr., USN, Program Executive, Aegis Ballistic Missile Defense, and Conrad J. Grant, Johns Hopkins University Applied Physics Laboratory, May 16, 2012, Washington, D.C.

[6]A partial profile of U.S. naval response to the Fukushima disaster—a combined earthquake, tsunami, and nuclear reactor catastrophe—in a coordinated effort known as Operation Tomodachi is found at http://www.nbr.org/research/activity.aspx?id=121. Accessed June 13, 2012.

Missions such as air defense and undersea warfare, which U.S. naval forces conduct in the open ocean and the littoral regions all have myriad entry points from which capability surprises can originate (land, air, space, and cyberspace). There are also accelerating new technological advancements globally, which again, singularly or in combination, can constitute the basis of a capability surprise.

Given the complexity of surprise, there is no simple way to guard against it. A number of explicit actions are needed. First and foremost, leadership must help others recognize the importance of understanding capability surprise and what it demands of U.S. naval forces, such as ensuring that organizations include preparation for and mitigation of surprise as one of their functions, including scanning and related activities, in order to discern potential surprises. Here, it is important that organizations are timely and diligent in examining the scope and seriousness of such surprises, and that they can identify other organizations that might be able to help anticipate, mitigate, or respond to these surprises.

A number of different kinds of surprise were discussed. One was an adversary's deployment of disruptive technologies against naval operations (such as specific "Zero Day" cyberoffense payloads). Another was the potential interruption of critical supply chains (such as those for rare earth elements). Yet another kind of surprise was the unfolding of geopolitical events (such as regional economic instability) that affect national security. The committee reviewed case studies of previous surprises and some of the circumstances leading up and surrounding them. Three of these are discussed next.

Surprise Scenarios

Three scenarios were selected that could be used to study capability surprises and, ultimately, to make recommendations:

- Scenario 1: Denial of space access;
- Scenario 2: An asymmetric engagement with complex use of cyber-methods in a naval context; and
- Scenario 3: A black swan event, to which the front-end scanning and prioritization framework for mitigating surprise is not applicable.

These three surprise scenarios are pertinent to U.S. naval forces and were used to carry out the tasks listed in the study's terms of reference. The committee selected them in part because each addresses an issue important to one of the three naval forces. Denial of space access is treated in a Navy context, social media manipulation is framed to apply in a Marine Corps environment, and domestic disaster relief is largely a Coast Guard mission. They are introduced below and more completely described in Appendix A. The scenarios are also used to illustrate key points throughout the chapters, with a summary of how they would be addressed via the proposed framework in Chapter 8.

Surprise Scenario 1 (An Intelligence-Inferred Surprise)

The committee considers the denial of space access to include the potential loss of space access due to antisatellite capabilities and electronic or optical countermeasures, including loss of intelligence, surveillance, and reconnaissance (ISR) feeds, communications, navigation (GPS), and timing (also GPS). The space-access scenario has been much discussed in the open media[7]not to mention that U.S. naval warfighting systems rely on positioning, navigation, and timing (PN&T). In essence, an adversary, in order to deny access to space assets of our forces, could employ the following measures against the space assets, either simultaneously or with unpredictable frequency, to render dependence on those assets unreliable:

- Jamming the GPS receivers of U.S. combatants or,
- Mounting cyberattacks on command and control (C2) centers and combatants,
- Jamming or dazzling surveillance sensors to obscure U.S. orbital ISR observations,
- Jamming communications reception by satellite receivers, or
- Kinetic engagement of orbital systems.

It was also recognized that cyberattacks or other interference in this scenario could originate from threats embedded in commercial off-the-shelf (COTS) hardware and software that is widely deployed in present naval systems and could render naval systems and networks inoperable at the critical moment of need. The reliance on certain widely used satellite communications operating in frequency bands that can be more readily jammed is a particular concern. Jamming of communications and denial-of-service attacks are clearly an intelligence-inferred surprise that can be mitigated by deploying alternative systems.[8]

Surprise Scenario 2 (A Disruptive Technology and Tactical Surprise)

The committee then considered an example of disruptive technology and tactical surprise—an asymmetric engagement with complex use of cybermethods.

---

[7]For example, see *Background Briefing on Air-Sea Battle by Defense Officials from the Pentagon*, News Transcript, U.S. Department of Defense, Office of the Assistant Secretary of Defense, November 9, 2011; available online at http://www.defense.gov/transcripts/transcript.aspx?transcriptid=4923, accessed May 9, 2012. See also Toshi Yoshihara and James R. Holmes, 2012, "Asymmetric Warfare, American Style," *U.S. Naval Institute Proceedings* 138(4):24-29; Andrew Erickson and Amy Chang, 2012, "China's Navigation in Space," *U.S. Naval Institute Proceedings* 138(4):42-47; and David Fulghum, 2012, "Under Siege: Foreign Countermeasures Proliferate as U.S. Electronic Warfare Programs Falter," *Aviation Week & Space Technology* 174(13):22-23.

[8]The issue of cyberdefense for U.S. naval forces will be covered more extensively in an upcoming NSB study, commissioned by the CNO, and anticipated to begin in 2013.

An adversary could use social media to gather a crowd that could place U.S. personnel and property at risk in foreign countries or threaten our domestic infrastructure. The potential implications of population unrest, spontaneous or induced, were explored. As happened in the Arab Spring, social media can be used to turn a local population against the United States and to facilitate the coordinated search and engagement of U.S. citizens and U.S. assets on the ground.[9] This is an event that may require a combination of tactics, techniques, and procedures (TTPs) and perhaps the creation of a new situation awareness capability, especially as it might involve naval ships and personnel or other U.S. naval resources operating in foreign ports. This scenario also lends itself to examining the complexities of attempted cybermanipulation of a crowd's mood and actions and provides a context for the potential engagement of on-the-ground and coastal operation.

Surprise Scenario 3 (A Black Swan Surprise)

Black swan surprises comprise a full range of unexpected events that could have a significant impact on the capabilities of U.S. naval forces. They might come from natural disasters (tsunamis, earthquakes, disease outbreak, and the like) or from decisions made in the face of political or economic exigencies. One national strategic decision was the recent decision to deploy the U.S. Coast Guard to remote areas of global conflict and, earlier, U.S. naval forces' provision of humanitarian aid and disaster relief (HA/DR) after the earthquake in Haiti in 2010.

Exemplars

In addition to various data-gathering activities and discussions that helped it formulate the three surprise scenarios described above, the committee was briefed on several programs that appear capable of timely anticipation and response to surprise. It decided to explore the three surprise scenarios and three of the programs, which it called "examplars,"[10] to help illuminate the following:

- Certain areas of potential surprise outside the mainstream acquisition programs that may impede anticipation and/or response;
- Successful principles and infrastructures that might be integrated into existing naval organizational structures and processes to deal with broader capability surprises;
- Structures and processes that could accommodate the three surprise scenarios that are currently unaddressed or underaddressed (space access, flash mob arranged for by social media, disaster response);

[9]Lisa Anderson. 2011. "Demystifying the Arab Spring," *Foreign Affairs* 90(3):2-7.

[10]Exemplars are programs that the committee thought were particularly promising.

- Capabilities, policies, and metrics that support successful structures and processes; and
- Changes to better prepare for capability surprise mounted against naval forces and to be more resilient to it.

The three exemplar programs chosen by the committee are (1) the Navy SSBN Security program,[11] (2) the Air Vehicle Survivability Evaluation program (Air Force red team),[12] and (3) the Aegis Ballistic Missile Defense (BMD) program, whose responsiveness was exemplified by the shooting down of the wayward National Reconnaissance Office (NRO) satellite in Operation Burnt Frost.[13] The principles and key ingredients used by each exemplar program to deal with potential capability surprise are similar: a stable program and infrastructure; a capability thread that includes research and technology development; modeling and simulation; expert staff; acquisition and industrial capability; testing infrastructure; and very visible senior leadership support and top cover. Several organizations, including the Marine Corps' Expeditionary Forces, the Coast Guard forces responsible for responding to natural disasters, the Office of the Secretary of Defense (OSD) Rapid Prototyping Office, and the Joint Interagency Task Force-South (JIATF-S) have developed a remarkable resilience that allows them to anticipate and respond to rapidly developing, on-the-ground needs.[14]

A unique fourth exemplar should be briefly mentioned: It is the ability of the highly adaptive and responsive special operational forces (SOF), in which Navy SEALs play an important role, to plan and induce surprise. Events revealed to the public during the past decade testify to how SOF can engage covert nonstate adversaries or apply novel surprise capabilities to degrade an adversary nation state's power. These forces are perhaps one of the best means for the United States to impose surprise on others.

---

[11]Stephen C. Schreppler, Andrew F. Slaterbeck, and CAPT Christopher J. Kaiser, USN, Office of the Chief of Naval Operations, N97, "SSBN Security Program Perspectives," presentation to the committee, April 12, 2012, Washington, D.C.

[12]Christopher Roeser, MIT Lincoln Laboratory, "Air Force Red Team Overview," presentation to the committee, May 16, 2012, Washington, D.C.

[13]RADM Brad Hicks, USN, Program Director, Aegis Ballistic Missile Defense, "Aegis Ballistic Missile Defense: Press Briefing, March 19, 2008," presented to the committee by RADM Joseph A. Horn, Jr., USN, Program Executive, Aegis Ballistic Missile Defense, and Conrad J. Grant, Johns Hopkins University Applied Physics Laboratory, May 16, 2012, Washington, D.C.

[14]Benjamin Riley, Principal Deputy, Deputy Assistant Secretary of Defense for Rapid Fielding, "Rapid Prototyping Perspectives," presentation to the committee, February 29, 2012, Washington, D.C, For additional information on the OSD Rapid Prototyping Office, see National Research Council, 2009, *Experimentation and Rapid Prototyping in Support of Counterterrorism,* National Academies Press, Washington, D.C.

CAPABILITY SURPRISE FRAMEWORK

With the above scenarios and exemplars in mind, and to guide the approach and understanding needed to address potential capability surprise for U.S. naval forces, the functional framework shown in Figure 1-1 was developed. It consists of six phases that can be aligned with the development functions, accountabilities, and principles observed in the exemplar programs.

The six phases—(1) Scanning and Awareness, (2) Assessing Surprise, (3) Prioritization, Option Development and Decision Formulation, (4) Resource and Transition Planning, (5) Implementation and Fielding, and (6) Force Response—are introduced briefly below and discussed in more detail in the following chapters.

- *Phase 1, Scanning and Awareness.* Involves scanning the horizon for potential technologies, technical applications, and operational behaviors that could

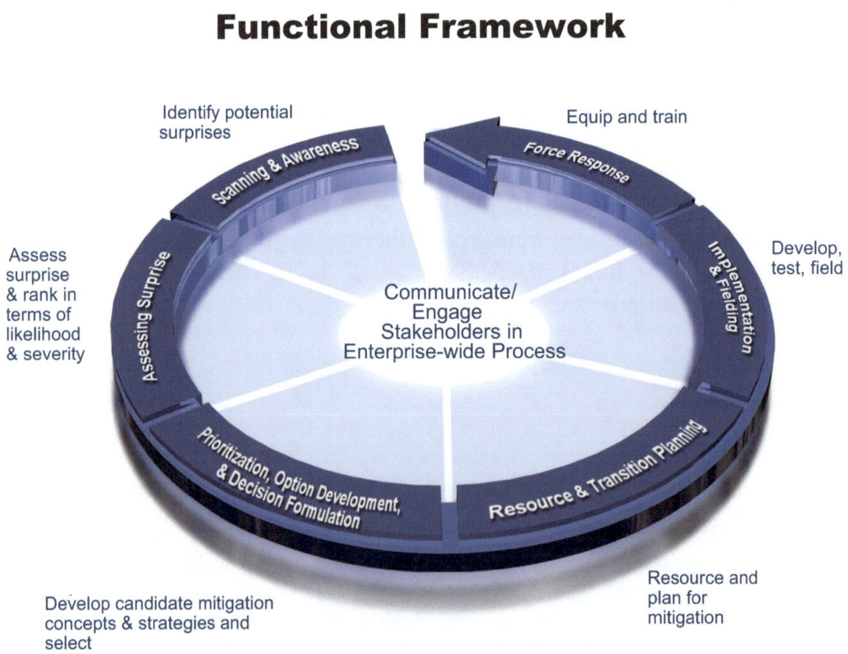

FIGURE 1-1 Six phases required for mitigating capability surprise. This is a continuous process in which each element informs the next. For example, force response adjustments may generate a loop-back in the process. Further reaction to a tactical or black swan surprise may enter one of the three right-most phases depending on the nature and timing of the required response. This framework can be used as a guide, with the understanding that surprises may occur at different points in the cycle.

cause surprise, which is defined here as a lack of preparedness or awareness to counter unexpected developments.

- *Phase 2, Assessing Surprise.* Includes key items like effective modeling, simulations, analysis, and red teaming. The somewhat overused term "red teaming" is applied here to emphasize the dynamic tension required of the operational, technical, and intelligence communities to discern potential negative impacts of surprise and to prioritize those that should be addressed in each time frame, from short term through long term.
- *Phase 3, Prioritization, Option Development, and Decision Formulation.* Includes concept development and evaluation to scope out possible solutions to the surprise risks identified in the preceding phase.
- *Phase 4, Resource and Transition Planning.* Involves identification of naval organizations to be assigned the task of resourcing and delivering the capability needed to address the surprise associated with the first three phases.
- *Phase 5, Implementation and Fielding.* Can take several forms depending on the results of the previous four phases:

—Development of new tactics, perhaps using existing assets or technologies in unexpected ways;

—Development of new variant capabilities within existing programs—for example, converting the software of a surface-to-air missile to a surface-to-surface missile, as was done with SM-1 in the 1970s;

—Rapid prototyping in order to field a few critical units as either sufficient to meet the need or as a stop-gap before an item can be produced and acquired by conventional means;

—Use of naval support centers to make changes to systems that are in service but out of production; and

—More aggressive use of Quick Reaction Capability (QRC) or other authorities.[15]

- *Phase 6, Force Response.* U.S. naval forces test operational capability, leveraging the U.S. naval test infrastructure; ensure training and proficiency; and determine the impact of a new capability on readiness in the face of surprise.

These six phases form the basic organizational structure for this report. Each of the Chapters 2 through 7 discusses one phase. The report's final chapter brings together the overall concept and highlights the key aspects of the proposed framework. It is important to recognize that these six phases are necessary to successfully anticipate or react to potential or real surprises. Specifically, the first four phases allow for the impact of a surprise scenario to be assessed so that it can be assigned a high or low priority. In Phase 4, several decision outcomes

---

[15]"QRC programs leverage DODI [Department of Defense Instruction] 5000.02 procedures and authorities to speed up the fielding of systems and capabilities to satisfy near-term urgent warfighting needs." See Air Force Instruction 63-114, January 4, 2011, *Quick Reaction Capability Process,* p. 5, para. 1.1.

are possible, including the development of new tactics. At this point, there is a natural tendency to implement a "quick and dirty" partial solution between Phases 4 and 5, and to ignore the residual risks to operations of such a solution. Accordingly, the adequacy of proposed responses should be assessed before any outcome emerges from Phase 5.

Phases 1 through 4 would help naval forces to anticipate intelligence-inferred surprises. Natural disasters (although not those on a black swan scale, such as the March 2011 Fukushima disaster) would enter the framework at Phase 5. Events in theater might require tactical or strategic operational adjustments in Phase 6, when the adequacy of proposed responses is assessed.

In addition to the Navy SEALs being part of SOF, naval forces often support SOF. The committee surmises that SOF and naval forces could figure out even more ways to collaborate to enhance both the mitigation of surprise (defensive) and the implementation of surprise (offensive). It would therefore appear highly advantageous to have SOF represented on the teams that address scanning and awareness and training and readiness, and possibly the other aspects as well of the surprise framework.

CROSSCUTTING ACTIVITIES

The committee recognizes that three activities pervade all the phases of the framework: (1) Modeling and Simulation, (2) Red Teaming, and (3) Research and Experimentation. It introduces these activities and discusses them where they first play a major role. Therefore, Modeling and Simulation and Red Teaming are introduced in Chapter 3, "Assessing Surprise." Research and Experimentation are discussed primarily in Chapters 2 (research) and 7 (experimentation) in the context of feasibility determination and resilience building, respectively. Figure 1-2 indicates the relationship between the three activities and the six phases of the framework.

OPTIONS FOR COORDINATING SURPRISE MITIGATION

The committee discussed in detail how to recommend leadership and associated activities that could enable our naval forces to implement the surprise mitigation framework. It first recognized that, besides the exemplar programs, virtually every development and acquisition program has some form of the following:

- Threat-response activity that includes projecting future threats,
- Development effort toward meeting the projected threats, and
- Continual assessment of existing and developmental system designs for potential vulnerabilities so that corrective measures can be addressed.

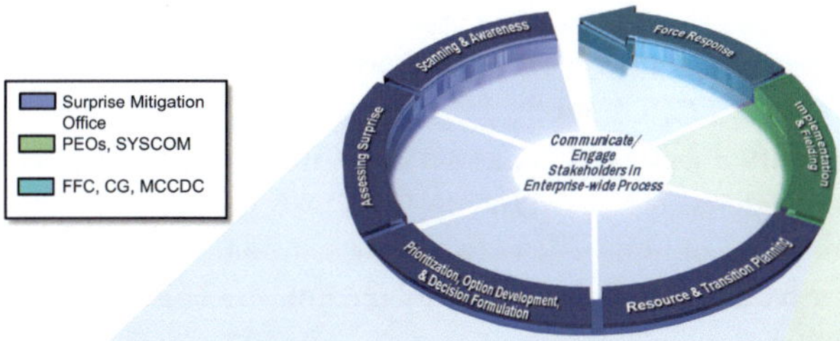

	Scanning & Awareness	Assessing Surprise	Prioritization, Option Development, & Decision Formulation	Resource & Transition Planning	Implementation & Planning	Force Response
Modeling & Simulation	Back-of-the-envelope calculations	Phenomenology models for feasibility System models for response Campaign models for impact	System models for trade-off analyses	Cost models Logistics planning models	System models for test planning and prediction	Verification and updating of models
Red Teaming	Experts debate risks and timing	Wargaming with cultural effects included	Red team concepts and capabilities	Review of planning	Red team test vignettes relative to threat	Exercise red team to challenge trainees
Research & Experimentation	Feasibility exploration of potential disruptive technology	Impact assessment via experiments	Validation via prototyping and testing	Risk reduction verification tests	Assess effectiveness via testing and tactics development	Operational readiness exercises and tactics experiments

FIGURE 1-2 Cross-cutting activities mapped to the functional framework.

The committee was therefore primarily concerned with those mission, or mission support, capabilities for which:

- There is presently no authorized office to ensure similar considerations,
- There are vulnerabilities that span a number of programs and for which there is no identified executive to coordinate the surprise mitigation,
- The surprises pop up as a result of emerging technologies or applications of existing technology whose potential impact on mission capabilities is presently ambiguous, and
- There is not sufficient attention being paid to identified risks.

The committee was especially mindful that in the present environment of limited resources and the need to consider affordability, the setting up of a new office would be considered unrealistic. On the other hand, the committee was keenly aware that it would be ineffective to just declare the need for the framework to be addressed and to state that the naval culture should consider potential surprises. This was the Navy's initial approach with FORCEnet, i.e., the realization of network-centric operations. While the Navy generally defined FORCEnet as the operational construct and architectural framework needed for network-centric operations, the construct and framework were ineffective by themselves. In short, additional resources and governance mechanisms, among other things, would be needed to realize FORCEnet.[16]

To meet the committee's obligations as set forth in the terms of reference, eight options were considered for coordinating and ensuring adequate attention to surprise mitigation. The options fell between the extremes of a completely new office on the one hand and the mere encouragement of a culture of surprise anticipation on the other.

Option 1: Incorporate into the Existing N9I Office

The N9I office ensures integration across programs; specifically, across warfare systems (expeditionary, surface, undersea warfare, and air warfare). Although its main role is integrated financial and budget management, the N9I also manages the Naval Integrated Fire Control-Counter Air (NIFC-CA) program. This program is particularly interesting because it requires technical and operational integration of sea surface, land, and airborne combatants (including Aegis, Army Aerostat, and E-2D) that share information over networks to engage airborne cruise missiles over terrain that is potentially beyond the line of sight of the firing units, using composite track and targeting data. The ability of N9I to coordinate programs is essential to mitigate surprises that are probably not addressed by a single program or capability upgrade. To extend the N9I office's mission to also address capability surprise mitigation seems synergistic, but it does introduce the need for additional staffing and coordination.

Option 2: Incorporate into the Existing N2/N6I Office

The N2/N6 information dominance role resonates with a number of potential surprises, including the scenarios (Appendix A) and exemplars (Appendix B) considered for space access and social networking. However, the N2/N6I already has a key role to play with respect to intelligence operations. Further, whereas the solutions to capability surprise threats may be implemented by N2/N6, the vulnerabilities leading to the need for mitigation of the loss of capability will likely apply to mission systems. For example, whereas alternatives to GPS and satellite communications might best be resourced by N2/N6, mitigating the impacts of the reduced availability of GPS and communications on a wide variety of combatants, commands, sensors, and weapons appears to go beyond the scope of N2/N6 and appears to be a better fit in N9I. The N2/N6, however, should be included in any arrangements made by the entity responsible for surprise mitigation.

[16]National Research Council. 2005. *FORCEnet: Implementation Strategy,* The National Academies Press, Washington, D.C.

Option 3: Establish a New Center of Excellence

One option considered was to declare an existing organization within the naval community as a center of excellence for surprise mitigation. The organization would be the one where technical capability is most closely aligned with the initial phases of the surprise mitigation framework—that is, it would best be able to interpret surprise risks by horizon scanning, red teaming, and analyzing the risks and options, and then to make recommendations. Potential centers of excellence that were considered included the Navy Warfare Development Command and the Naval Postgraduate School. There were a number of drawbacks to this approach, however. First, the need for deep technical expertise beyond that available at such centers requires the ability to identify and the authority to marshal experts from across the naval community, such as technical specialists, especially those with validated system and physics models, as well as cultural experts. Furthermore, as mentioned in Chapter 3, there is a tension between the need to continually turn over, and refresh, expertise and the need to maintain a permanent expert staff. This is not to say that the aforementioned organizations, and others, do not have a key role to play. It is just that the particular combination of qualities—authoritative assembly, temporary duties, and prioritized resourcing—does not exist in a single, established technical center.

Option 4: Assign to the Office of Naval Research

Since the Office of Naval Research (ONR) is recognized as key to organizing research horizon search and assessment activities, through leadership from ONR-Global in collaboration with the Office of Naval Intelligence and others, and since ONR is expected to lead in the formation of critical technical experiments and prototyping, perhaps it could be assigned responsibility for coordinating surprise mitigation for naval forces. On the other hand, the committee believes that such an office will need the robust involvement of the operational community as well as the technical community, which would make the Office of the Chief of Naval Operations (OPNAV) a more desirable place from which to coordinate surprise mitigation.

Option 5: Create a Rapid Acquisition Office with PEO

Over time, program executive offices (PEOs) have been established or re-chartered to embrace new programs. For example, the PEO for Theater Air Defense of the 1990s was expanded to become the PEO for Integrated Warfare Systems (PEO-IWS), with additional acquisition programs and missions in surface and undersea warfare and tactical command and control. Whereas such PEOs can be instrumental in addressing the broader problems of emerging needs, they do not necessarily address the emerging cross-mission and pop-up threats in a manner that also considers operational options.

Option 6: Create a New OPNAV Office

The committee observed several recent lessons, including the N00X office (the Naval Warfare Integration Office in OPNAV), which was established several years ago and more recently disestablished. Although the N00X office, while it existed, provided useful broad context for potential naval capability needs and deficiencies, it was deemed insufficiently resourced, staffed, and authorized to be effective. The committee thought it best that an existing authoritative office in N9, with a compatible charter but access to resources, be expanded in scope and authority so as to provide greater impact and response.

Option 7: Delegate to a Surprise Office in the Office of the Secretary of Defense

OSD established a Strategic Capabilities Office last year featuring

> a shift of $60 million to support the stand-up of the new strategic capabilities office charged with analyzing emerging threats, developing innovative and architecture-level concepts, intelligence concepts, red teaming and conducting demonstrations of disruptive technologies.[17]

This is a positive DOD initiative to enhance preparedness against surprise and appears to incorporate some of the essence of the first several phases of the framework presented and recommended in this report. One could argue that with the OSD office now in place, the need for a naval surprise mitigation office is obviated. However, while the committee considers the OSD office to be a focus of collaboration across the Services, the office would not be expected to focus on predominately naval surprise issues. Furthermore, mitigation of surprises directed against our naval forces must be implemented by the naval organizations themselves, especially in the case of rapid fielding, acquisition system upgrades, and the development of tactics, techniques, and procedures. Still, the support of the OSD office should make it unnecessary to concentrate on Joint services' response to surprise, thereby freeing the recommended naval surprise mitigation office to focus more on naval-specific concerns.

Option 8: Incorporate into Existing DASN RDT&E

The committee also considered whether it would be appropriate for the Deputy Assistant Secretary of the Navy for Research, Development, Testing and Evaluation (DASN RDT&E) to establish a capability surprise mitigation office. Such an option could provide advantages in terms of rapid prototyping develop-

[17]Christopher J. Castelli. 2013. "DOD Seeks $11.98 Billion for Science and Technology in FY-14 Budget," *Inside Defense.Com,* April 9.

ment and modification of acquisition systems. Moreover, the committee believes that the DASN RDT&E should play a key role in ensuring that appropriate technical responses to potential threats and surprises are considered in program design reviews and upgrade planning (the committee's belief is embodied in its recommendation that DASN RDT&E serve as the Chief Technology Officer. However, the committee believes that responding operationally to potential surprises should be the fundamental driver and that inputs and preparedness efforts begin and end with the operational commands. Accordingly, the committee believes that such an office is best suited to reside in OPNAV, where it will have greater access to operational communities.

In summary, the committee considered the N9I office within OPNAV to be the most likely existing organization from which to coordinate surprise mitigation for naval forces in concert with joint forces and national assets. However, the committee considered that it would be presumptuous to recommend that office if the CNO determines, based on plans and considerations of which the committee is not aware, that another office is more appropriate. Therefore, the committee recommends the establishment of a surprise mitigation office, considers N9I the most likely organization that could grow into this role, but understands that the CNO may identify a more appropriate entity to take on the surprise mitigation coordination role according to the framework presented and recommended in this report. Regardless of who is assigned responsibility, an office with sufficient authority and connectivity, as defined by the recommended framework, is necessary to effectively address surprise. As pointed out directly by the Defense Science Board, "Rarely is there a case of true surprise. Post mortems almost always identify that someone has provided warning, but that warning was not heeded."[18]

LEADERSHIP

The committee acknowledged the challenges and complexities faced by naval forces in dealing with capability surprise, as exemplified in the three surprise scenarios and the three exemplars described earlier and in Appendixes A and B. In each of these scenarios and exemplars, the various stakeholders (e.g., operational, intelligence, technical, and acquisition related) should be involved to bring awareness of the potential vulnerability posed by capability surprise. Likewise, different entities should be responsible for prioritizing, resourcing, exercising, and developing TTPs against such scenarios. For example, entities ranging from the Atlantic and Pacific fleets to the Office of Naval Intelligence to the Office of Naval Research may be involved in scanning for potential surprise. Laboratories, naval operating forces, the Marine Corps Combat Development Command (MCCDC), the Navy's OPNAV N2/N6 and N9 organizations, and the PEOs have

[18]Defense Science Board. 2009. *Report of the Defense Science Board 2008 Summer Study on Capability Surprise*, Volume I: Main Report, Office of the Under Secretary of Defense for Acquisition, Technology and Logistics, Washington, D.C., September, p. xii.

key roles to play in prioritizing and developing responses and ensuring readiness. While many stakeholders are involved, there does not appear to be a designated lead working across the Navy to ensure recognition of these potential capability surprises and also to ensure the required integration and prioritization of efforts to help mitigate their negative impact. There does not seem to be a supporting infrastructure or lead integrating authority to work through the complexities that cut across various naval authorities.[19] Not just new or emerging technologies must be scanned and addressed: Scanning activities should also search for the use of existing technologies and capabilities in unforeseen ways. Other countries' exercising, doctrine, publications, and technologies must be scanned. This leads to the major finding and recommendation of this report, Finding 1 and Recommendation 1, which tie the naval organizational framework to the six-phase functional framework for capability surprise that is introduced above and discussed in more detail in the subsequent chapters of this report.

FINDING AND RECOMMENDATION

Finding 1: Capability surprise is both inevitable and inherently complex, and it requires U.S. naval forces to engage in a broad spectrum of issues, from horizon scanning to red teaming to experimentation and rapid prototyping, to exercising, fielding, and training. While there are a few exemplar organizations in the Navy, Marine Corps, and Coast Guard as well as the Department of Defense that effectively work on capability surprise, there is neither an overall framework for nor a clear delineation of U.S. naval forces responsibilities.[20]

Recommendation 1: The Chief of Naval Operations (CNO), the Commandant of the Marine Corps (CMC), and the Commandant of the Coast Guard (CCG) should establish a common framework for U.S. naval forces to address capability surprise, as shown in Figure 1-3.

The CNO, in concert with the CMC and the CCG, should establish and fund a surprise mitigation office to serve as the executive agent for addressing capability surprise for U.S. naval forces. Specifically, this office should comprise a set of organic operational, technical, assessment, and intelligence-oriented staff and draw input and analysis with a global, mul-

[19]The potential impact of a recently announced OPNAV structural reorganization, creating the N9 as a single office to oversee warfighting programs, is a step toward providing structure that may help mitigate capability surprise. The impact of this new structure will become apparent over time. For additional information on this realignment, see Navy Office of Information, "CNO Realigns OPNAV Staff," March 3, 2012. Available at http://www.navy.mil/search/display.asp?story_id=65845. Accessed May 24, 2012.

[20]Some exemplars identified by the committee include the Navy's SSBN Security program, the Air Vehicle Survivability Evaluation Program (Air Force Red Team), and the Aegis Ballistic Missile Defense program (in response to Operation Burnt Frost).

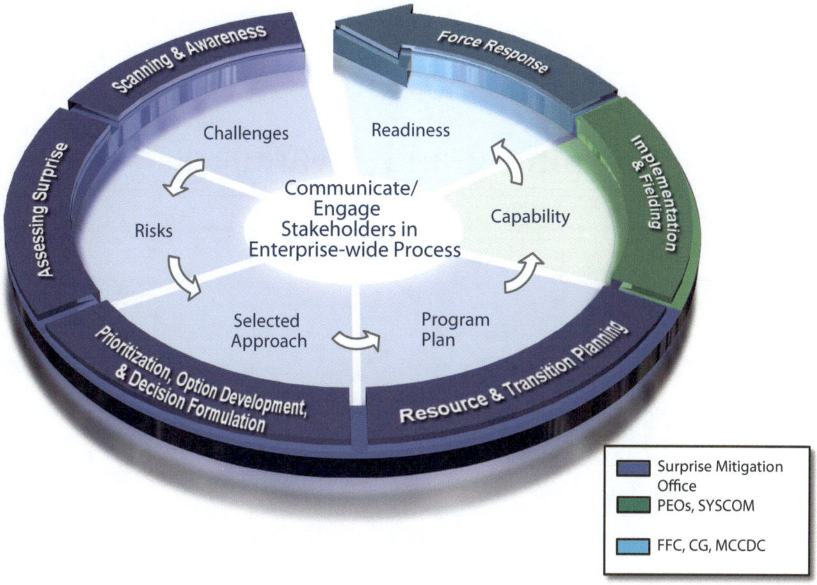

FIGURE 1-3 Recommended framework for addressing capability surprise.

ticultural perspective from a multitude of communities: operational; intelligence; acquisition, research, and development; system commands; program executive offices; war colleges; military fellows; government laboratories; Federally Funded Research and Development Centers; university-affiliated research centers; industry; and academia. To ensure that U.S. naval forces are proactive in developing and anticipating surprise capabilities it is recommended that representatives from Special Operations Command (SOCOM) be an integral part of this office. Based on the CNO's decision above on which option to pursue (see Table S-1), the office should support the timely transition of surprise capabilities to the appropriate organizations of primary responsibility and monitor the transition to, and effectiveness within, operational forces.

2

Scanning and Awareness

BACKGROUND

Studies by the Defense Science Board (DSB)[1] and the Naval Research Advisory Committee (NRAC)[2] have slightly different recommendations for surveillance, and the report of the Center for a New American Security (CNAS)[3] has prescriptions to better account for predictive failure. Taken together, these documents form a reasonable basis for the naval scanning and awareness approach that will lead to preparation or avoidance strategies in succeeding steps. The committee specifically notes that the *Report of the Defense Science Board,* rather than focusing on avoidance, focuses on preparation (in which surveillance plays a major role), flexibility, and rapid response across the domains of operations, technology, and rapid acquisition. In contrast, *Disruptive Commercial Technologies* emphasizes surveillance.

Figure 2-1 presents a continuous cycle of scanning and sifting for early detection of new adversary technologies, tactics, and operational concepts; a capability projection of how those could translate into new military capabilities; a net assessment of what those adversary capabilities might mean at an operational or strategic level; an options analysis to determine the effectiveness, cost, and schedule of alternative means of response; and a decision package that provides

[1]Defense Science Board. 2009. *Report of the Defense Science Board 2008 Summer Study on Capability Surprise*, Volume 1: Main Report, Office of the Under Secretary of the Office of Defense for Acquisition, Technology and Logistics, Washington, D.C., September.

[2]Naval Research Advisory Committee. 2008. *Disruptive Commercial Technologies,* Office of the Assistant Secretary of the Navy for Research, Development and Acquisition, Washington, D.C., June 26.

[3]Richard J. Danzig. 2011. *Driving in the Dark: Ten Propositions About Prediction and National Security,* Center for a New American Security, Washington, D.C., October.

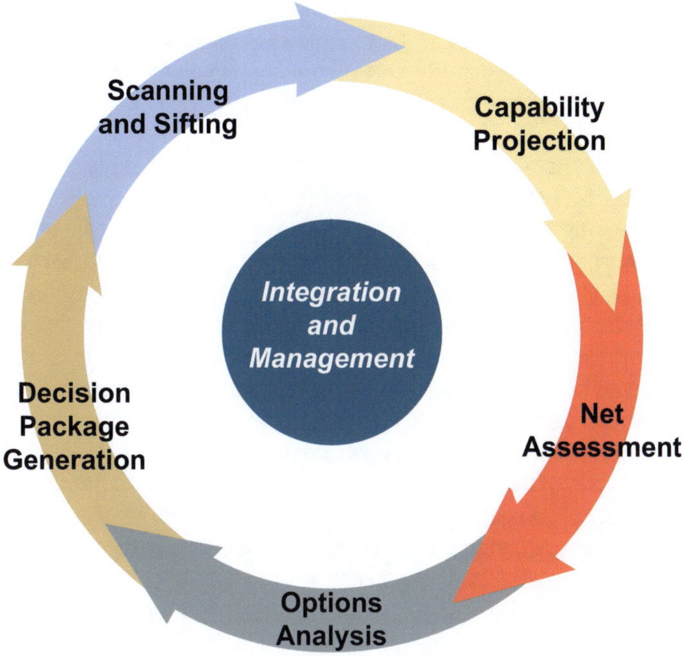

FIGURE 2-1 The surprise management cycle. SOURCE: Defense Science Board. 2009. *Report of the Defense Science Board 2008 Summer Study on Capability Surprise*, Volume I: Main Report, Office of the Under Secretary of Defense for Acquisition, Technology and Logistics, Washington, D.C., September.

a comparative assessment and actions plan for those alternative responses. The proposed naval process includes scanning and awareness as approximately the scanning and sifting and capability projection elements of the DSB cycle. The committee's main departure from the DSB process is focusing on the naval organization and on the implementation of and training for surprise mitigation.

It is also the case that the NRAC report used the "red cell" concept:

> The NRAC Disruptive Commercial Technologies Study Panel undertook an unusual side excursion and conducted a Commercial Red Cell Demonstration to investigate the potential of creative people with World Wide Web access to produce new ideas, anecdotally determining what capacity a Red Cell might have to defeat key USMC capabilities or gaps. This experiment tested our hypothesis concerning the potential of small, Internet-enabled groups to interfere with key USMC capabilities. We reached out to the Hollywood creative community for two reasons:

–One of our panel members had previous experience with Hollywood concept development efforts, and was confident that the results would be positive.
–As a shared value, people in the Hollywood community are accustomed to the idea of *ad hoc* groups tackling novel tasks in highly focused efforts.[4]

The committee covers the red cell approach in Chapter 3, "Assessing Surprise."

The Office of Naval Research-Global (ONR-G) already undertakes a form of horizon scanning for research. As will be described next, its role could be expanded to cover not only emerging research results but also emerging commercial technologies and some types of enigmas. As identified in the submarine security program, a form of scanning known as "enigmas" is an activity that examines which technologies and physical phenomena could be exploited even if they are not yet observed in the horizon scanning. Sometimes deployed forces, from fleet commands to Special Operations Forces (SOF) on the ground, observe a surprise during their horizon scanning activities. Sometimes those observing the emergence of a surprise, SOF in particular, are best equipped to respond to or mitigate the surprise, if they are properly trained for such contingencies. Instructing naval forces in contingency planning has proven to be a versatile response to surprise that will be addressed in a later chapter.

Scanning and sifting require multiple intelligence modes—financial, open-source, human, and other clandestine means. The projection of technological maturity and capability requires technically qualified support and an ability to adopt the mindset of an adversary. Surveillance may require an organization to "own" this responsibility lest it be neglected, and it should be a standing organizational element because surveillance activities need to be ongoing throughout the lifetime of any program. Note that ONR, by virtue of "owning" both ONR-G and ONR S&T programs involving experts from laboratories, universities, and industry throughout the United States can leverage both.

The prescriptive remedies of the CNAS report should be considered.[5] Many of these appear to be like procurement actions but can be strongly connected to experience of people using the product(s) in the field. This means these remedies should be included explicitly or integrated into field use, as methods for evaluating (or even planning) the required adaptations and informing of future revisions. The CNAS report does not strongly reflect surveillance, although its call for "nurturing diversity and creating competition" in order to "produce a valuable range of potential responses when unpredictable challenges and difficulties arise" fits in somewhat with surveillance concepts.[6] Scanning and sifting activity from

---

[4]Naval Research Advisory Committee. 2008. *Disruptive Commercial Technologies,* Office of the Assistant Secretary of the Navy for Research, Development and Acquisition, Washington, D.C., June 26, p. 5.

[5]Richard J. Danzig. 2011. *Driving in the Dark: Ten Propositions About Prediction and National Security,* Center for a New American Security, Washington, D.C., October.

[6]Ibid., p. 6.

the DSB report is more relevant. It is interesting that the NRAC study does not explicitly refer to the surveillance process, in spite of the fact that its central example (of a brainstorming effort by outsiders, with spectacular results) was used explicitly as a "scanning" process.

EXAMPLES FROM THE COMMERCIAL AND ACADEMIC SECTORS

At a session of the TTI/Vanguard Advanced Technology Conference Series in Seattle on December 6, 2012, the Chief Technology Officer (CTO) of Intel, Justin Rattner, spoke on rethinking industrial research on the 21st century. His main theme was based on a remark by Alan Kay, "The best way to predict the future is to invent it." He went on to discuss how Intel, as well as 12 benchmarked peer multinational industrial research laboratories, explores future opportunities and threats with the intention of creating its own future. Intel invests 50 percent of its research on exploratory areas, and the other 50 percent on areas that align with the present business. Whereas the business-aligned developments focus on improvements to the present product lines, the exploration topics can be far afield of the present business. Intel focuses on exploratory areas of promise to position the company for its own breakthroughs, thereby "inventing the future." If breakthroughs come from elsewhere, Intel, with its related exploration investments, might still be positioned not far behind others to remain a serious competitor for the new area. Because explorations are generally beyond the collective expertise of Intel staff, or perhaps because an insufficient core of expertise exists within the company, Intel and its benchmarked peers tend to go outside of the organization to partner with universities in open research. Breakthroughs that occur are generally developed into products, not within Intel but as joint ventures outside Intel's product lines. This approach appears to align with the messages in *Innovator's Dilemma*[7] and *The Other Side of Innovation.*[8] Technology explorations beyond present product, or operational, lines must be protected and nurtured in neutral territory to provide sufficient expertise and prevent the natural tendency of a product line from resisting potentially competing innovations.

Another example of similar thinking is the proposed approach for more flexible federal government policy preparation. A National Defense University (NDU) report, *Anticipatory Governance,*[9] explores how to provide the capability in the U.S. government to explore potential global developments, from black swans through more slowly evolving situations, so that U.S. policies and contingency

---

[7]Clayton M. Christensen. 1997. *The Innovator's Dilemma: When New Technologies Cause Great Firms to Fail,* Harvard Business Review Press, Boston, Mass.

[8]Vijay Govindarajan and Christopher Trimble. 2010. *The Other Side of Innovation: Solving the Execution Challenge,* Harvard Business Review Press, Boston, Mass.

[9]Leon S. Fuerth with Evan M.H. Faber. 2012. *Anticipatory Governance Practical Upgrades: Equipping the Executive Branch to Cope with Increasing Speed and Complexity of Major Challenges,* National Defense University, Washington, D.C.

plans are available to address the increasingly complex and dynamically changing international landscape in a timely manner. For this objective a set of agile activities and organizational adjustments is recommended. The approach focuses on the Executive Branch and includes the involvement of technical, policy, cultural, and strategy experts, many from academia, to explore potential future developments and the adequacies of present and potential policies.

Central to changing the government culture to better accept (and not dilute) such potentially outside-the-box findings, a "foresight fusion cell" would operate inside the White House as "top cover" as the cell performs horizon searching and would explore the implications of its finding. Recommended changes in policy are then initiated by the Executive Branch. Some process streamlining, changes of organizational scope, and improved inter-organizational interfaces are also recommended for government entities that would participate in the finalization and implementation of new or changed policies. The combination of executive backing and systems engineering-oriented interface modifications to these other entities is expected to eventually shift government culture toward broader acceptance of timely policy innovations.

The proposed role of the White House foresight fusion cell and modifications to other government structures is analogous to the activity and organization of Intel's exploration investments to create the future. A takeaway from these examples is that truly outside the box developments—that is, those with the most potential for disruptive surprise—are likely best pursued outside the naval organization, in academia and small business ventures, even if sponsored by interested naval program entities. This is the approach ONR-G has continued to pursue, identifying promising science and technologies and resourcing academia across the world. As a corollary, moving such emerging capabilities from the laboratories to the naval establishment too soon could lead to their rejection or dilution of their potential (competing) impact by the more conventional naval organization.

Surprises arising from the commercial sector can have significant impact on the U.S. government and Department of Defense. These surprises can be due to the unexpected use of, and innovation based on, commercial technologies—for example, the emergence of improvised explosive devices (IEDs) during the Iraq war[10]—or from adversarial exploitation of vulnerabilities in the commercial infrastructure on which the U.S. government relies (e.g., the break-in to RSA's networks that compromised the SecureID two-factor authentication token widely used within government).[11] The wide availability of technology combined with the ability to innovate, particularly in the cyberdomain, is a game-changer. Cybercriminals, in particular, are using techniques that not so long ago required na-

[10]J. Nicholas Hoover. 2012. "NSA Chief: China Behind RSA Attacks," *InformationWeek Government*, March 27. Available at http://www.informationweek.com/news/government/security/232700341. Accessed March 30, 2012.

[11]See an entry title "SecurID" at https://en.wikipedia.org/wiki/SecurID. Accessed March 30, 2012.

tion state–level resources (e.g., botnets, social engineering and big-data mining, and multinational bases and agents) to gain illicit access to financial resources.[12]

THE SCANNING AND AWARENESS APPROACH

Recognizing and preparing for the different kinds of capability surprise requires advanced studies and war games conducted with open-minded experts across a wide range of fields. Their results need to be continually brought to the attention of the most senior leadership.[13] Specifically, experts on capability surprises need to be leading red teams that challenge "conventional" mission executions. Further, outsiders such as college students need to be used to bring new perspectives to what might be done to achieve capabilities surprises. As has been recognized on several occasions, students quickly came up with various unexpected counters to potential capability surprises or thought of new surprises that were fast and easy to implement.

Because of the very diverse nature of the sources of, especially, longer-term capability surprises, many organizations must be actively involved. To make this cross-organizational activity work, processes must be instituted that organize and assign responsibilities to each of the participants and that permit aggregating the diverse information to enable passing it to the "Assessment Processes" described in Chapter 3. Further, there must be processes to provide feedback to all of the parties such that they can be aware of the larger picture and also to help them focus on areas designated as higher priorities. Lastly, there must be a process for periodic senior-level briefings and involvement. This will provide awareness for their decision-making processes and permit them to anticipate and perhaps modify the organization's direction as early as possible based on potential surprises. Further, it will be necessary for the senior personnel to become involved so as to resolve issues arising across the various organizations due to priority or resource conflicts. It will permit review of this broad cross-organization for performance and budget adjustments as the effort progresses. This is important as each individual organization may be reluctant to adequately budget for efforts that support the broader organization. Senior-level people will have the authority to modify as needed the various components as the efforts evolve. Last, and perhaps most important, the "senior spotlight" on this issue will ensure energetic actions by all parties to do their part to avert altogether or at least mitigate capability surprise.

---

[12]Melissa E. Hathaway. 2011. "Taking a Byte Out of CyberCrime," Science, Technology, and Public Policy Program, Belfer Center for Science and International Affairs, Kennedy School, Harvard, Cambridge, Mass. Available at http://belfercenter.ksg.harvard.edu/files/byte-out-of-cybercrime-hathaway-oct-2011.pdf. Accessed April 4, 2012.

[13]A single interaction is not sufficient, as seen with Einstein having to twice appeal to President Franklin D. Roosevelt on the potential of nuclear weapons.

Assigning Responsibilities and Developing Processes

The great diversity of scanning and maintaining awareness requires a structured process having three elements: (1) stakeholders, (2) participants and performers, and (3) activities whose outputs are provided to the Assessing Surprise functions described in Chapter 3 and as appropriate to the later chapters. A virtual organization can be created to leverage resources across the many organizations that take part in this structured process. The ONR-G, as will be described later, can be the foundation for this virtual capability surprise structure. Current ONR-G activities, which include surveying for S&T, could be broadened to include the interests of the other stakeholders and maintain central control of the data for all of the organizations. Further, ONR-G could develop and maintain a prioritized list of potential capabilities surprises according to their technology readiness or expected maturity time lines. The head of ONR-G could provide the list and updates on a monthly (if not a more dynamic) basis to the Chief of Naval Research (CNR) and the surprise mitigation office as well as to other relevant organizations. This list together with the integrated views of the Office of the Chief of Naval Operations (OPNAV) could also be provided to the Chief of Naval Operations (CNO) every quarter. ONR-G would ensure that all of the inputs from the other stakeholders, such as the Office of Naval Intelligence (ONI) and the Deputy Director of National Intelligence (DNI) for S&T, are captured and feedback is provided. In addition to coordination with ONI, ONR-G would interface as broadly as reasonable with the intelligence community and, specifically, with the Open Source Center. The fleet forces, as the deployed naval presence abroad, can be a valuable observer of worldwide trade and tactics training. Therefore the Service commands should organize to provide input and operational assessment expertise for this ONR-G-led activity. Finally, ONR-G will be empowered and resourced to task certain other stakeholders and S&T organizations, such as laboratories and universities, providing information and/or expert assessments as needed.

Office of Naval Research-Global

ONR-G today has a well-structured program able to survey and assess S&T around the globe (Europe, Asia, and South America). It has in place many of the processes that can be applied to the other stakeholder activities. ONR-G has ties to many countries and their respective technical organizations, including its U.S. Air Force and Army counterparts, which could have additional insights that might be relevant to the naval forces. In addition, because of its very close ties to the other ONR activities and to various domain experts, ONR-G enjoys regular in-depth reviews of various technical subjects with a range of interested and knowledgeable parties. The resulting information is distilled and provided to senior-level persons. The effort and reporting are integrated with additional scanning products and provided to senior management quarterly and archived in a structured manner for follow-on assessment and prioritization, as described in Chapter 3. This database will serve as a long-term, updatable repository of information as other elements of a potential capability surprise are aggregated and assessed. Whereas ONR-G has a network of international research ties, ONR itself is well connected to the network of U.S. research organizations. Thus, ONR is truly both national and international in its connectedness.

A key function of ONR-G is not only to scan global research activities for important emerging science and technology but also to actively foster research by providing funding, and, as the opportunities arise, promoting research collaborations with U.S. universities and research centers. Further, if research in an area is showing signs of a potential breakthrough or of a game-changing application, ONR-G can continue the international research, increase funding for it, and perhaps add a U.S. collaborator, or the broader ONR might establish a related research activity in the United States. This research could be extended to critical experiments or to the collection of scientific data needed to better assess feasibility and potential impact. It is important to ensure that the United States gain a hands-on understanding of potentially fruitful research that would allow it to anticipate and mitigate a surprise or to develop its own surprise. By having a domestic research capability, the time between awareness and reaction can be shortened because the learning curve has already occurred. In summary, the "scanning and awareness" phase should include active U.S. research and experimentation to best assess risks and potential.

Office of Naval Intelligence in Coordination with Naval Scientific and Technical Intelligence Officers

Coupled closely with the ONR-G and other stakeholders, intelligence gathering can identify capability surprise candidates and, more importantly, can be requested to investigate, in more depth, things that might be indicators of a capability surprise. Other members of the intelligence community—particularly the intelligence agencies of the Office of the Secretary of Defense (OSD) such as the DNI and the intelligence offices of the other Services—can also provide inputs, preferably coordinated by ONR-G or by the ONI on behalf of ONR-G.

Offshore Scanning by Offices

As they carry out their role of rebalancing toward Asia, the Pacific Fleet (PACFLT) staff—in particular the fleet science advisors and operational units—can provide a wide range of insights across various technology and security domains. The same is true for the Atlantic Fleet (LANTFLT) with its vantage from the Commander, U.S. Naval Forces Europe (COMNAVEUR), to monitor activities in the European, African, and South American theaters. While these organizations today have responsibilities to report on and to interact in their communities, the scanning and awareness network led by ONR-G can correlate and analyze on a broader scale and make requests to organizations for additional information. Further, based on insights gained by capability surprise organizations, requests or alerts may go to a central point in the fleet. It is anticipated that Commander, Fleet Forces Command (FFC), and PACFLT, and perhaps certain numbered fleets, will have horizon-scanning offices. Further, counterpart activities, perhaps as offices, could be enlisted in the effort, including the Coast Guard Force Readiness Command (FORCECOM) and the Marine Corps Combat and Development Command (MCCDC).

Chief Technology Officer

Scanning the horizon for technologies that are threats as well as opportunities should go beyond just the mitigation of surprise. Although ONR works to disseminate and apply emerging technologies, the inherent "valley of death" between emerging technologies and applications remains. Although the proposed surprise mitigation framework is intended to bridge this gap, a senior naval technologist leader would not only provide the additional coverage at the top, but could also serve to introduce technologies into the fleet in an effective and efficient manner. Therefore, the Navy should appoint a CTO to ensure adequate bridging of the gap between technology's emergence and its implementation. The CTO would serve as an advisor to the surprise mitigation office and also would ensure appropriate technology insertion at the appropriate risk level throughout the naval capability development organization, primarily by having a decisional role in major milestone reviews. Because he now has the most closely aligned role, the Deputy Assistant Secretary of the Navy for Research, Development, Testing and Evaluation (DASN RDT&E) is in the best position to be appointed as the naval CTO. Care should be taken in establishing this new position to respect its historical context.

Others

An ongoing process aims to provide warfighters with rapid reaction support for their urgent operational needs that are identified. The naval forces seniors from both the Secretary of the Navy and the CNO are part of the senior integration group (SIG) that reviews and resources critical needs. The first priority of SIG is surprise. Establishing a close working relationship with this group will ensure that naval interests are supported.

To play their respective roles in national security, the national laboratories, academia, industry, and the intelligence and other communities naturally turn to ONR-G as a connection point. ONR-G either already enjoys such connections or could leverage the connections of its ONR parent.

When a "flash response" is called for, metaorganizations provide an overarching leadership framework affording guidance, direction, and momentum across organizational lines. For the naval forces to react quickly to capability surprises, a set of contingency connections should be developed that includes reaction approaches, medical responses, media and legal support to articulate some areas, and rapid preparedness for further attacks. By bringing primary and alternative contacts, communications backups, and related systems together in metaorganizational packages, the effects of capability surprises can be mitigated.

The committee held two teleconference meetings with the staff and leaders from the Commander, Pacific Fleet (COMPACFLT). The members of the committee, who included very senior (retired) fleet commanders, recognize that the integrated horizon scanning demands observations of the activities of other naval forces and even commercial traffic in the theater. It was also pointed out that these in-theater observations may generally be shared with other commands but not with OPNAV, S&T, or the acquisition establishments except for local command use in planning and intelligence backup.

FINDING AND RECOMMENDATION

Finding 2: The Office of Naval Research-Global (ONR-G) is focused on scanning at the 6.1 and 6.2 levels; the Office of Naval Intelligence (ONI) on technical intelligence, primarily systems in development, testing, or operational exercises (6.3-6.7 levels); and fleet intelligence on observed operational behavior. However, there is no integrated, comprehensive scanning that also explores the linking of these observations and the emergence of consumer technologies of potential impact. While there are usually early indicators of potential capability surprises, there does not appear to be a coordinated means for U.S. naval forces to explicitly scan the horizon for such indicators; to capture, retain, and vet such indicators with relevant organizations; and, ultimately, to inform senior leadership of potential capability surprises. Such a coordinated means would need to scan, recognize, categorize, analyze, and report technical and/or operational surprises on a global basis.

Recommendation 2: Using existing fleet resources, the Chief of Naval Operations should enlist the support of the combatant commanders and their naval component commands in order to scan, recognize, capture, and report potential capability surprises outside the continental United States (OCONUS). Most notably, the Commander, U.S. Naval Forces Europe and Africa, and the Commander, U.S. Pacific Fleet, should establish comparable counterparts within their respective staffs to the surprise mitigation office referred to in Recommendation 1.

To further aid in identifying potential capability surprises—technological, operational, and/or otherwise—the Assistant Secretary of the Navy for Research, Development and Acquisition (ASN RDA) should (1) appoint the Deputy Assistant Secretary of the Navy for Research, Development, Testing and Evaluation (DASN RDT&E) as the Navy Department's Chief Technology Officer, with primary responsibility for providing technical advice for all phases of the framework for the surprise mitigation office in Recommendation 1 and (2) direct the Chief of Naval Research to establish a "virtual" scanning and awareness structure led by the Office of Naval Research-Global, engaging the technical, intelligence, and operational communities in order to systematically scan the horizon, maintain awareness, and conduct technology readiness assessments for both the CNO and the surprise mitigation office, as called for in Recommendation 1.

3

Assessing Surprise

METHODS TO ASSESS AND ANALYZE SURPRISE

Intelligence-inferred surprise comes not from a lack of knowledge but from an inability to adopt countermeasures to meet a new threat. Even disruptive technologies and tactical surprises should typically have some leading indications, most often recognized in hindsight.

To evaluate responsiveness to threats and develop new tactics to respond to surprise, one must practice in an environment that closely mirrors reality. Accordingly, naval forces often exercise red teams and rely on anticipatory modeling and analysis to predict surprise, identify vulnerabilities, and develop countermeasures, either by rapidly fielding existing response technologies or, in extreme cases, engaging the acquisition process to build new naval capabilities.

In the course of this study, a number of communities were heard from that have successfully implemented red teaming,[1] modeling, and analysis. Through use of separate teams within the maritime force structure or leveraging not-for-profit labs (government and industry) to evaluate threats to national security, the Department of Defense (DOD) has established a formal mechanism to evaluate and respond to technology surprise *in specific areas*. A number of entities were heard from in detail that illustrate successful independent evaluation: Air Force Red Team program, the SSBN Security program, and the Aegis Ballistic Missile Defense (BMD) program (in response to Operation Burnt Frost), to name

[1]Red teaming and many of the other tools discussed in this report have applications in many of the recommended phases for addressing surprise. They are not necessarily discussed in detail every time they may be relevant.

three. The activities of each of these exemplars are provided in Appendix B to this report.

In each of these examples, the red teams have been granted independence in assessing vulnerabilities and evaluating threat responses. Each entity also has access to a strong base of expertise available to brainstorm vulnerabilities and solutions. Technical subject matter experts (SMEs) in academia and industry are engaged as necessary; they perform detailed analysis, use their imagination, or brainstorm on a particularly challenging problem. In the SSBN Security program, for example, the diverse SME team is preplanned, extensible through outreach, and explicitly identified as the "Friends of SSBN" network; it is a standing team, ready to be called on to respond to surprise issues.

It is also noted that in order to identify threats and anticipate surprise, red teams perform modeling, simulation, and analysis at three levels of fidelity: (1) campaign-level modeling validated through (2) system-of-systems simulation made realistic by (3) high-fidelity physics-based models. Successful implementation of this multitiered modeling involves an ability to leverage existing simulations that are being developed in the national laboratories and industry, often by individuals in the SME networks. Running exercises and threat scenarios through this three-tiered modeling and analysis capability will identify potential threats, allow for response evaluation, and identify potential vulnerabilities and risk. Subsequently, in-depth vulnerability analysis (including precise evaluation of algorithms, software, hardware, or system performance issues) has proven essential to determining the impact of a threat and the necessary response.

In some cases, this response will require a change to existing assets or acquisition of a new technology. Therefore, red teams are able to recommend and/or deploy solutions to the field as necessary.

The methodology for assessing and responding to surprise that is used by the SSBN security red team serves as an excellent representative approach (see Figure 3-1) for evaluating vulnerabilities in large programs of record. As an independent group that seeks to challenge the organization in order to improve effectiveness, the SSBN Security program leverages simulation, modeling, and analysis to assess risks to submarine security and recommends mitigation strategies. Similar success has been noted in the approaches used by the Air Vehicle Survivability Evaluation Program. Each of these exemplars leverages modeling and analysis tools in conjunction with a network of experts to expose bias, offer critical review, model vulnerabilities, and demonstrate alternative ways to respond to surprise.

U.S. naval leadership should leverage the approaches used by the three exemplar organizations (SSBN security, Air Force red team, Aegis BMD's Operation Burnt Frost) to further anticipate, model, and simulate both intelligence-inferred and potential disruptive surprise. The hallmarks of these successful approaches are as follows:

- Senior leadership support (commonly termed "top cover");
- Team independence;
- Access to a strong base of cross-disciplinary technical and operational expertise with corresponding models and simulations that have been verified and improved via test results;
- An ability to identify threats through campaign-level modeling, system-of-systems simulation, and high-fidelity physics-based models (see next section for further explanation);
- Precise vulnerability modeling and analysis capability;
- Mechanism to recommend and/or deploy solutions as necessary;
- Adequate, steady funding; and
- Focus on a particular mission.

The committee recognizes that naval forces face a wider spectrum of challenges than just those addressed by these exemplars. Nevertheless, the approaches used by them can serve as a model for successfully addressing capability surprise in other complex mission areas.

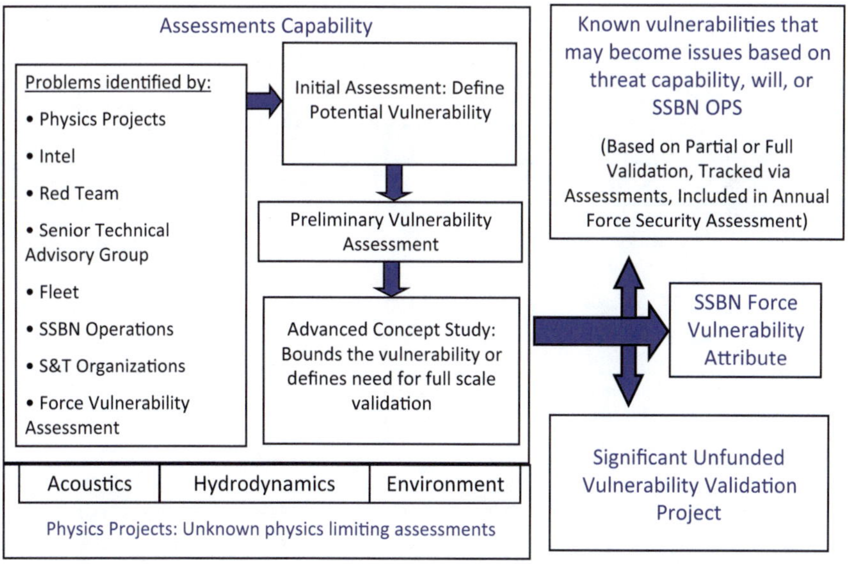

FIGURE 3-1 Methodology used by the SSBN Security program. SOURCE: Stephen C. Schreppler, Andrew F. Slaterbeck, and CAPT Christopher J. Kaiser, USN, OPNAV, N97, "SSBN Security Program Perspectives," presentation to the committee, April 12, 2012, Washington, D.C.

SYSTEM-OF-SYSTEMS MODELING AND SIMULATION FOR EXPERIMENTATION AND RISK REDUCTION

Assessing potential surprise and its impact on the naval forces requires input from and interaction among a diverse set of communities, including naval forces, the intelligence community, the defense industry, national laboratories, universities, and commercial industry. In some sense, these communities can themselves be viewed as "sensors" that collect and process information that is relevant to detecting potential surprise, assessing its likely impact, and formulating measures to deal with it. However, each community has domains in which its sensors are more effective and domains where they are less effective. Figure 3-2 attempts to provide some sense of domains where the strengths of the various communities reside.

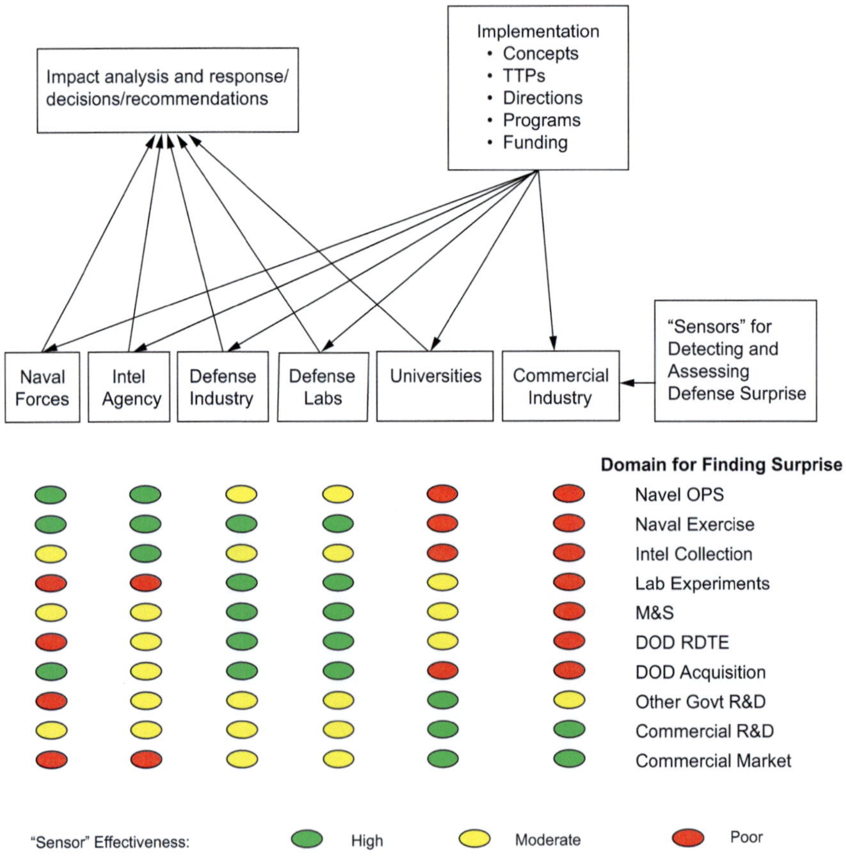

FIGURE 3-2 Domains of surprise "sensor" effectiveness.

No single community has strong "sensor" capability in all the domains that can uncover potential surprises. It is therefore necessary to take advantage of the collective strength of the relevant communities so as to gain a more coherent and authoritative understanding of the potential for surprise, analyze its potential impact, and create and implement solutions. In other words, this assembly of sensors must itself be viewed as a system. The complexity of this system is clear from Figure 3-2, as is the need for sophisticated tools to exploit it. Modeling and simulation (M&S) is a natural tool for dealing with this type of complexity and has become an integral part of determining the capabilities and vulnerabilities of naval systems and assessing the potential for surprise. The overall undertaking is vast, and the models employed cover many different disciplines: basic physics, engineering design for development of platforms and weapon systems, mission planning, training, intelligence gathering and interpretation, conduct of actual military operations, and the subsequent evaluation of their outcomes. Clearly, a complete discussion of M&S in support of the naval forces would not be practical. However, as an attempt at illustration, the brief discussion in this section will focus on the role of M&S using first principles.

To achieve the M&S capability it needs, the naval scientific and technical community is organized into various discipline areas for the purpose of developing and maintaining the SMEs and facilities needed to support naval missions. The disciplines include platform (ship, aircraft, spacecraft, etc.) design and construction, radar systems and technology, acoustic systems and technology, missile systems and technology, electronic warfare systems and technology, communication systems and technology, space systems and technology, electronics, materials, chemistry, environmental (ocean, atmosphere, and space) science and technology, and more. The overall community spans the government and private sectors. It necessarily includes not only theoretical modeling and analysis capability but also evaluated results of well-designed and instrumented tests to validate the M&S and ensure their improvement as greater systems and phenomenology understanding emerges from the tests and experiments. Each of the SME communities is responsible for developing the levels of models needed to meet its responsibilities for contributing to naval superiority. The various SME communities must interact with one another. For example, the placement of radar on a ship influences ship design and vice versa. Electronic warfare systems need input from and also influence radar systems and communication systems. Radar systems, acoustic systems, electronic warfare systems, and communication systems all need input from the appropriate environmental communities. The required working-level ties among the various SME communities needed for model development are maintained in each SME community—the sponsors of the SME communities require this. This rather loose federation has proven to be effective for advancing M&S in all the communities. The process becomes much more formal when a decision is made to proceed with development of a particular weapon system or with conduct of a fleet exercise. In either case, coordinated and specific input from several SME communities would be required to perform the M&S needed for that development or exercise or when an effort is initiated to assess the potential for surprise in a particular area. At that point, an overall program management structure is usually employed, an organization is designated to coordinate the overall M&S program, and the appropriate contractual obligations are put in place. This requires significant interaction among the relevant SME communities, the naval staff, the appropriate acquisition offices, the operational test and evaluation (OT&E) community, and others as necessary.

Modeling, simulation, and analysis tools are critical for the development and deployment of naval platforms, systems, logistics, and training. These tools, when properly leveraged by a network of SMEs in academia, industry, laboratories, and service colleges, form a strong framework within which to anticipate and respond to surprise.

TECHNOLOGY-FOCUSED VULNERABILITY STUDY GROUPS

There exists a set of known surprises that are of immense concern to naval forces. Many of these surprises have surfaced as lessons learned from humanitarian assistance and disaster relief (HA/DR) efforts, from discovery during fleet exercises, or in the course of regular mission operations. The committee heard common concerns from Coast Guard, Navy, and Marine Corps leadership about port security, mine warfare, cyberthreats, use of commercial shipping for hazardous materials, and small vessel threats. It recognizes that government agencies are working to characterize these concerns and understand their potential impact. However, inconsistencies were found in the way the results of these efforts were being applied in exercises and red team activities. In some sense, because everyone is concerned about these threats, no one organization appears to be tasked with truly characterizing them and therefore with identifying threat response and mitigation strategies for widespread use. In other words, it appears difficult to focus on disruptive threats within the existing infrastructure and processes.

Vulnerability assessments for critical technology infrastructure have so far been based on small exercises and the like, and the effects must be better quantified, modeled, and/or characterized in order to be leveraged more effectively into the operation of fleet forces. For example, threats to precision navigation and timing sources, the potential for cyberattacks embedded within navy weapon systems, or vulnerability of the intelligence, surveillance, and reconnaisance (ISR) system could significantly impact naval operations; however, little evidence was found that the element of surprise has consistently been incorporated into fleet exercises and training.

Similarly, while known potential threats to U.S. interests (including those that threaten the homeland)—for example, semisubmersibles, small vessels, and the mining of U.S. ports—are acknowledged and characterized as disruptive surprise issues and potentially catastrophic, there appears to be a lack of training and exercises and insufficient preparation by U.S. naval forces in responding to these threats and in developing a concept of operations for response using existing capabilities in the event these threats materialize.

It is encouraging, however, that a significant body of knowledge exists within the wider DOD community in the form of M&S resources and a network of SMEs familiar with these subjects at various levels of fidelity. However, the disconnect between experts capable of better modeling and characterizing the threats and the fleet forces charged with responding to new threats leaves the Navy vulnerable to surprise.

The recommended surprise mitigation office could continually identify and prioritize vulnerable operational or technological areas that are not owned by a single mission. In each of these vulnerability areas, the office should establish a threat study group (TSG) to further characterize, quantify, and model the threat. A given TSG would leverage existing resources (modeling, simulation, and analysis tools used by a network of SMEs in academia, industry, laboratories, and service colleges) to characterize the risk so that it will be useful for campaign-level modeling and fleet exercises.

IMPROVING RED TEAMS THROUGH NONTRADITIONAL PERSPECTIVE

A lack of appreciation for other cultures is a recognized weakness in U.S. intelligence and defense planning. Compared to other nationalities, Americans on average speak fewer foreign languages, travel infrequently beyond our nation's borders, and lack a working understanding of the cultural nuances that could result in surprise. There is overall, a great tendency to look at how an adversary will behave from an American standpoint. Sadly, however, when assessing the potential for an adversary to introduce surprise, it is the adversary's cultural values, societal norms, and geopolitical priorities that must be taken into account. Naval forces must find a way to incorporate nontraditional perspectives into exercises, red team activities, and training efforts. The question is how to best approach this challenge from the standpoint of both human resources and technology infrastructure. History has numerous examples—for example, kamikazes, the Maginot Line, and suicide bombers—of not anticipating a surprise because of flawed cultural perspectives.

Throughout this study, red teams were heard from that were assembled to evaluate weapon systems performance, analyze specific technology vulnerabilities, and run naval exercises given various adversarial scenarios. These red teams leveraged a strong system-of-systems M&S capability that focused on the physical systems that naval forces would employ in any warfighting activity. However, there is little evidence of the systematic use of human social, cultural, and behavioral (HSCB) modeling to inform red team activities or exercises. It also found no consistent attempt to capture and incorporate nontraditional perspectives for these exercises from personnel with broader cultural backgrounds.

There appear to be major shortfalls in constructing red teams that can simulate or represent the thinking of adversaries across global cultures. As our understanding of potential surprise matures, it is important that independent red teams are assembled with the appropriate balance of skills and HSCB consideration to ensure an understanding from military and cultural perspectives that may not exist within the core organizations.

Fortunately, DOD recognizes the need for diverse perspectives and cultural understanding and has invested in HSCB research in order to introduce the human dimension into defense planning. In the course of this study, the committee found a number of examples of research in HSCB modeling that illustrate the potential for evaluating surprise taking into account cultural differences in a controlled training environment. For example, the Office of Naval Research has invested in an HSCB modeling program to build a knowledge base and create training capacity to understand, predict, and shape human behavior across global cultures. This program seeks to understand the HSCB factors that influence behavior at individual, group, and societal levels. In doing so, researchers are developing computational M&S capabilities, visual analytical toolsets, and mission-rehearsal systems that will give naval forces the ability to come up with a culturally sensitive forecast.

The committee recognizes that understanding the totality of human behavior is beyond the scope of HSCB and that in order to be effective for military planning, HSCB research must focus on the role of the military in the context of government actions and on behavioral understanding required to specify data and models relevant to the military missions.

In that light, various projects of the Institute for Creative Technologies (ICT) were also reviewed. Established in 1999 with a multiyear contract from the U.S. Army, ICT is a multidisciplinary research institute at the University of Southern California focused on exploring and expanding how people engage with computers, through virtual characters, video games, and simulated scenarios. ICT develops advanced immersive technologies to create human synthetic experiences that are so compelling the participants will react as if they are real. Herein lies a potential for anticipating and assessing surprise. Given promising quantitative interpretations of the qualitative findings of social psychology, military doctrine, and proven mathematical modeling techniques, ICT has developed a framework, PsychSim, that is sufficiently realistic to train users in cultural awareness, battle space preparation, and mission operations. In the safety of a virtual world, trainees interact with virtual humans and experience situations and dilemmas that they are likely to face in the real world. Such interaction drives outcomes based on choices that are made in this immersive environment.

The PsychSim framework is explained as follows:

> [Designed] to explore how individuals and groups interact and how those interactions can be influenced[,] PsychSim allows an end-user to quickly construct a social scenario, where a diverse set of entities, either groups or individuals, interact and communicate among themselves. Each entity has its own goals, relationships (e.g., friendship, hostility, authority) with other entities, private beliefs and mental models about other entities. The simulation tool generates the behavior for these entities and provides explanations of the result in terms of each entity's goals and beliefs. A user can play different roles by specifying actions or messages for any entity to perform. Alternatively, the simulation itself can perturb the scenario to provide a range of possible behaviors that can identify critical sensitivities of the behavior to deviations (e.g., modified goals, relationships, or mental models). A central aspect of the PsychSim design is that agents have fully specified decision-theoretic models of others. Such quantitative recursive models give PsychSim a powerful mechanism to model a range of factors in a principled way.[2]

Given the technology maturity of HSCB research for use in military applications, the committee proposes that select HSCB learning tools be introduced into the system-of-systems modeling capability that naval forces may leverage to evaluate potential surprise. It notes that the Navy has already invested in this technology through the Immersive Naval Officer Training System (INOTS) program. Targeting leadership as well as basic counseling for junior leaders in the Navy, "the INOTS experience incorporates a virtual human, classroom response technology, and real-time data tracking tools to support the instruction, practice, and assessment of interpersonal communication skills."[3] The Joint Improvised Explosive Device Defeat Organization (JIEDDO) has also employed this technology in counter-IED training systems involving video narrative, immersive environments, and geospecific multiplayer gaming scenarios. The Dismounted Interactive Counter-IED Environment for Training (DICE-T) system sends trainees on various interactive missions that emphasize critical components of dismounted patrol: planning a route, executing a patrol, and countering threats, and mission debrief or after-action review. The game scenarios represent real-world dismounted patrol situations. The DICE-T system teaches novices to think like experts before they are deployed.

These examples are cited as evidence that HSCB research has been transformed into useful tools for military planning. Given the growing need to understand human terrain and cultural differences in future military operations, the committee suggests that future red team participants could benefit from these

---

[2]Stacy C. Marsella, David V. Pynadath, and Stephen J. Read. 2004. "PsychSim: Agent-Based Modeling of Social Interactions and Influence," *Proceedings of the International Conference on Cognitive Modeling,* Pittsburgh, Penn. Available at http://people.ict.usc.edu/~marsella/publications/iccm04.pdf. Accessed February 6, 2013.

[3]Description available at http://ict.usc.edu/prototypes/inots/. Accessed July 10, 2013.

immersive training tools, which would accelerate their understanding of the military operating environment.

U.S. naval leadership should take steps to ensure that red teams, provided they have sufficient independence, expertise, and a fresh influx of participants, depart radically from traditional thinking in order to prepare forces for combat and have them to develop new tactics. Efforts must be made to diversify the composition of teams to better represent the adversary's thinking. It would be desirable for red teams to vary in terms of culture, ethnicity, and international experience. They should have multiservice, multigenerational, multidisciplinary, and independent backgrounds and should include nonmilitary, business/commercial, and academic sector members.

The committee recognizes that ONR has made significant progress by bringing together sociologists, anthropologists, psychologists, engineers, computer scientists, warfighters, and analysts to develop an influence operations portfolio for U.S. naval forces. However, the full integration of these HSCB models with the more traditional military models will require further advances in modeling the whole of society. Figure 3-3 illustrates the spectrum of modeling capability that is required to tackle the new challenges faced by the defense community. On the left side of the scale, one sees the traditional defense modeling capability. Based on known mathematical approaches, the Navy's simulation enterprise as described earlier in this report is based on the laws of physics and weapon system performance parameters validated through testing and operational use. On the right side of the scale, one sees the social science models, based largely on heuristics. Both modeling environments should be part of military assessment activities going forward.

As military forces are increasingly used as an integral component of overall political operations, red teams will need to bridge this gap, learning from HSCB models and using this perspective to inform military operations.

The committee recalls the words of Secretary of Defense Gates: "No one should ever neglect the psychological, cultural, political, and human dimensions

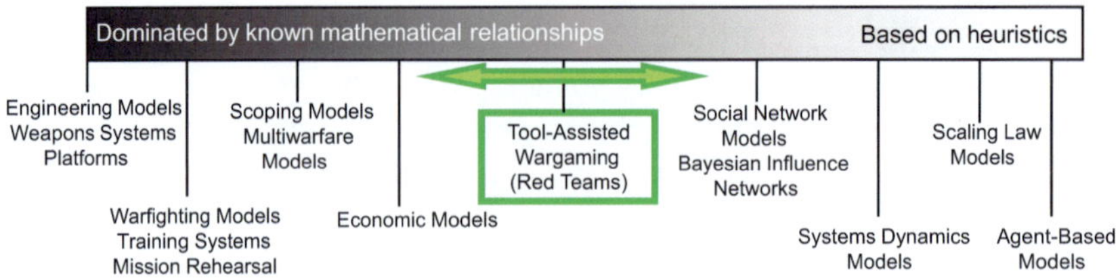

FIGURE 3-3 Defense systems modeling domain. SOURCE: Adapted from S.K. Numrich and Andreas Tolk, 2010, "Challenges for Human, Social, Cultural and Behavioral Modeling," *SCS M&S Magazine,* January.

of warfare. War is inevitably tragic, inefficient, and uncertain, and it is important to be skeptical of systems analyses, computer models, game theories, or doctrines that suggest otherwise."[4] This message continues to hold true.

MODELING AND RED TEAMING OPPORTUNITIES IN THE COMMITTEE DEFINED SCENARIOS

Modeling and Simulation in the Context of Space Systems

Over the years the U.S. Navy has developed ever greater use of and dependence on space-based assets. For example, the Global Positioning System (GPS) is now used and depended on for ship navigation, tracking, situational awareness, cartography, time synchronization, precision weapon guidance, force protection, command and control and logistics management, among other things. As another example, the recent launch of MUOS-1 continues the evolution of SATCOM into the regime of mobile users as they traverse the battle space. The use of space-based surveillance assets for battle planning, execution, and damage assessment continues to increase. All of the above are further enhanced by rapid developments in cyber capabilities. Collectively, these developments make space assets a significant force multiplier. As with all force multipliers, the use of and dependence on space assets is accompanied by vulnerabilities. These can range from a minor degradation of capabilities to a significant or even a total loss of that force multiplier. M&S provides critical tools for understanding the various vulnerabilities and for suggesting and assessing approaches to ameliorate or overcome the vulnerabilities.

It is helpful to have some understanding of how M&S plays in the space infrastructure. In a simplistic sense, one can break the problem into several categories: the spacecraft themselves; the physical environment in which the spacecraft sit; the propagation links that exist among the spacecraft and the various entities from which they receive or to which they transmit information; and the properties of the entities that transmit information to or receive information from spacecraft. The latter entities can be in space, in the atmosphere, on land, on the sea, or under the sea. The resulting systems are so complex as to themselves require significant levels of M&S in design, operation, and maintenance and in the response to their degradation.

Modern spacecraft typically consist of a platform or bus on which are installed sensors, processors, and communication equipment. Each of these areas requires extensive M&S in the design of the spacecraft, in its subsequent operation, and in the response to the various stimuli to which the operational spacecraft is subjected, manmade or natural. Spacecraft and their associated models are

---

[4]Robert M. Gates. 2009. "A Balanced Strategy: Reprogramming the Pentagon for a New Age," *Foreign Affairs* 88(1):28-40.

tested extensively on Earth. No spacecraft of any consequence is constructed without the involvement of a variety of simulation communities. These communities will generally follow the spacecraft through its operational life and play a key role in resolving problems that might arise with the spacecraft. The associated M&S is generally done using a "hardware in the loop" approach, where model output is used to stimulate the spacecraft and the spacecraft response is then used to stimulate the M&S. The same is true for the ground control systems that control the spacecraft. This type of M&S can be characterized as electrical, mechanical, and product assurance. A more detailed but still high-level discussion of this type of M&S can be found at the European Space Agency Website.[5] This type of M&S and the associated communities work at the state of the art and are an integral part of the international spacecraft program.

M&S of the environment in which a spacecraft sits is critical to its routine operation and to managing the spacecraft when it is subject to anomalies (natural or manmade). The M&S of spacecraft environment and associated natural anomalies (solar flares, energetic particle events, and the like) is covered under the rubric "space weather." Because of its broad military and commercial impact, there exists a National Space Weather Program, the details of which can be found at the National Space Weather Program Portal.[6] The DOD component is discussed in the *National Space Weather Program: The Implementation Plan* (2nd Edition).[7] This program encompasses monitoring, modeling, and predicting the space environment and its impact on SATCOM, GPS positioning and timing, and spacecraft. It pulls together all of the relevant U.S. M&S communities and represents the state of the art regarding the natural environment and its impact on space systems. However, there may be some concern regarding the M&S situation as it relates to high-altitude nuclear explosions. This will be commented on below in the context of GPS issues.

M&S related to the communication links among spacecraft and associated receivers and transmitters is a well-developed field. In a simplistic sense, this M&S attempts to predict the characteristics of the signal received by an antenna to the signal transmitted by a spacecraft, ground station, or other transmitter. It is generally referred to as link analysis and deals with matters such as system noise, environmental noise, atmospheric attenuation refraction and scattering, and sources of signal interference. The M&S tools are highly developed and the simulation community works at the state of the art. However, as with all M&S, the output of a simulation is no better than the input, which, as can be seen from the topics addressed by this M&S, can be subject to high levels of uncertainty or ignorance. As a result, this M&S becomes an essential tool in reconciling a

[5]Available at http://www.esa.int/Our_Activities/Space_Engineering. Accessed February 6, 2013.

[6]Available at http://www.swpc.noaa.gov/portal. Accessed February 6, 2013.

[7]National Space Weather Program Council. 2000. *The National Space Weather Program: The Implementation Plan*, 2nd edition, Office of the Federal Coordinator for Meteorology, Washington, D.C., July. Available at http://www.ofcm.gov/nswp-ip/pdf/nswpip.pdf. Accessed February 6, 2013.

spacecraft's expected performance with its observed performance. Understanding the essential interplay between expectations and observations is paramount in space systems (or any complex system) development. Designers have a tendency to understate matters that might reduce system performance, and simulators can only simulate based on their knowledge at the time. It is the reconciliation process that ultimately determines the system performance. The GPS is a good example of this interplay as it has occurred from its conception to its present status.

The GPS ranks among the great scientific and technical achievements of the twentieth century. It has changed the way that the world works, including the way that the Navy works. Its improvements are so seductive that its impact has penetrated far beyond that envisioned by its inventors and at a speed that is difficult to comprehend. It is quite likely that its impact on naval operations is not fully understood. Perhaps an interesting M&S undertaking might be to compile the full penetration of GPS dependence into areas that the naval forces depend on and to assess the broad impact on naval operations that the penetration implies. The results might be surprising.

One fact that is now well accepted is the relative ease of jamming GPS receivers owing to the low broadcast power of the GPS system. This creates a grave concern due to the ever deepening penetration of GPS-based technology into the military and commercial domains. One approach to dealing with this concern has been the attempt to significantly reduce the vulnerability of the GPS approach to navigation and time transfer. It has resulted in a several orders-of-magnitude improvement in jamming rejection over that available in the original GPS implementation. The M&S tools discussed above played an important role in this undertaking. One of the outputs of this activity has been the advancement of controlled reception pattern arrays (an approach to electronic beam forming and null forming to maximize GPS signal and minimize jamming signal) to the point where they are sufficiently reduced in size that they may be considered for a broad array of platforms. However, even with these advances, further improvements in antijam capability are needed, and other approaches should be considered to reduce cost. M&S tools will be essential to making progress and trade-offs regarding this much-needed capability.

Another approach to dealing with the jamming vulnerability has been to examine alternatives to GPS. The Office of Naval Research, for example, has funded a diversified program with the objective of identifying technologies that might form the basis of alternatives to GPS that could be viable for the naval mission. This program has advanced technologies in the areas of precision celestial navigation, inertial navigation based on fiber optics, microelectromechanical systems (MEMS), and quantum mechanical approaches and the development of small atomic clocks. It may be that several of these developments have advanced to a point where they by themselves or in some combination with one another could demonstrate sufficient promise as to warrant more aggressive development. The employment of the M&S tools analogous to those mentioned above may provide a cost-effective approach to make an initial assessment in this area.

A third approach has been to consider alternative fallbacks in the event of a significant degradation or loss of GPS capability. One such proposal has been to reestablish the now disabled LORAN navigation system but using today's technology. Such an effort is under way in the United Kingdom and Europe. That system is designated eLoran. The objective of this undertaking is as follows:

> The GLAs have been providing their prototype eLoran trial service since 2007. eLoran satisfies the requirements of the international maritime community for a digital e-Navigation future that specifies an independent, dissimilar and complementary backup for GNSS. eLoran allows users to retain GNSS-levels of navigational safety even when satellite services are disrupted.[8]

While this approach has obvious limitations regarding the U.S. Navy mission, there is certainly some intersection with the Navy's broad interest in reliable navigation and time transfer. M&S could be helpful in identifying the extent of the intersection.

A concern was mentioned above regarding M&S capabilities related to high-altitude nuclear explosions in the ionosphere. While such explosions are unlikely, they must be considered as a possibility with large consequences. The M&S concern relates to the ability to realistically model the geographically large ionospheric modification that will persist for hours after the explosion. Low earth orbit (LEO) spacecraft that survive the explosion may pass over this disturbed region. Such plasmas have a tendency to become highly structured (striate). Furthermore, it is known from natural events (generally referred to as Spread F) that the GPS signal can undergo significant phase fluctuations when it propagates through such plasmas. This can lead to deep fading of the signal at the GPS receiver causing obvious problems (such as loss of lock-on). The level of M&S sophistication needed to simulate the nuclear-explosion-induced plasma is high, and while it once existed, it is not clear that it still exists. Since M&S is the only way to quantify this problem, it would be a good idea to ensure the availability of the needed expertise.

There are many other aspects of the role of M&S in military space systems that could be discussed. However, it should be clear that its role has been and is essential. Furthermore, it is clear that M&S will play a key role in resolving the GPS vulnerabilities that confront the Navy and that there are potentially viable approaches to that end.

---

[8]Research & Radionavigation Directorate, General Lighthouse Authorities. 2013. "eLoran." Available at http://www.gla-rrnav.org/radionavigation/eloran/index.html. Accessed February 6, 2013.

Potential Use of Modeling and Simulation in the Social Media Manipulation Scenario

Only recently has the world seen the power of virtual communities influenced or called to action by social media. As outlined in the Crowd-Sourcing via Social Media Scenario (see Appendix A), which is about a fictitious place called Provencia, a well-planned Internet-based propaganda campaign can poison the hearts and minds of a population, escalating and organizing discontent to the point where lives are lost and military intervention is required. In this scenario, social media are an important means for our adversaries to rapidly change the game for U.S. interests. To operate effectively in this sort of complex, public opinion-laced scenario, U.S naval forces would likely leverage red teaming and M&S to prepare for both kinetic and nonkinetic response. Since kinetic systems M&S traditional red teaming has long been utilized throughout the Navy's system-of-systems enterprise, the committee next briefly explores how HSCB and social media modeling could enhance current M&S environments and better prepare naval forces for a scenario such as the one faced in fictitious Provencia.

Social Network Modeling and Influence Operations

Imagine a modeling environment and open simulation framework that allows naval forces to rapidly assemble a reasonable representation of the Provencia situation, including (1) forecasting and predictive models of human behavior for Provencia's general population, easily influenced activists, and Freedom for Provencia (FFP) renegades and (2) social media feeds to serve as a catalyst for these human behavior models. The simulation interface could allow military mission planners to create news reports or social media feeds that would then be parsed by the simulation software for keywords used to potentially enrage the population and introduce new behaviors into the Provencia environment model. Using this crowd-sourced, behavioral predictor, military planners could then evaluate the potential impact of kinetic and nonkinetic responses to the uprising.

What if the Provencia situation became a prolonged challenge for U.S. interests? At some point, the HSCB framework could be leveraged to model social networks within Provencia, understand the network's nodes and tipping points, and evaluate or influence operations scenarios. What would it take to convert public opinion in favor of the United States? Which social media (blogs, postings, etc.) would have the most impact and should therefore be exploited to address the situation? These are among the questions that could be considered through the use of a robust HSCB modeling tool. It is important to note that the usefulness of an HSCB modeling tool would depend on how well the culture is understood.

Leveraging HSCB Models into Red Teams and Wargaming

Ideally, one could merge physics-based systems models with those based on social science to create a comprehensive simulation environment to be used in defense operations planning. However, at this time, it would not be that simple. Engineering models for weapon systems and platforms are firmly based on mathematical principles, whereas social science models are based largely on heuristics and do not lend themselves to traditional approaches for validation, verification, and accreditation. Therefore, for the near future, the committee suggests developing a mechanism to leverage what is learned through the use of these HSCB modeling tools and then using it for a broader red team and wargame scenario. For the situation in Provencia, a red team should be prepped for the situation using the human behavior modeling framework and then employed in a wargaming experience that has been carefully orchestrated to include social media–based upsets and nonkinetic response. Red teams should be varied in culture and international makeup and have multiservice, multigenerational, and multidisciplinary personnel, including those well versed in the social sciences.

SUMMARY

Ultimately, the United States must train and equip combat-ready naval forces capable of deterring aggression, maintaining freedom of the seas, and, if deterrence fails, winning wars. As the world becomes more connected and threats have become more obscure, a new surprise mitigation office on behalf of stakeholders in the Navy Fleet Forces Command (FFC)-Navy Warfare Development Command (NWDC), USCG Force Readiness Command (FORCECOM), and Marine Corps Combat Development Command (MCCDC) must consider new perspectives in assessing surprise. This activity must take full advantage of experts in industry, academia, the national laboratories, and government to identify and characterize risk, consider solutions, develop mitigation plans, and deploy change.

As was observed in the exemplar programs and as has been the general experience of committee members concerning the acquisition programs, the technical intelligence community is generally responsive to requests and tasking concerning potential technical threats to mission systems. However, for those systems that represent the infrastructure support to mission systems, such as GPS, ISR, and communications, the committee did not observe that any mission organization was tasking, reviewing, and assessing threats against infrastructure systems; consequently, some organizations tended to assume their availability. At any rate, the committee observes that if a responsible organization were taking threats to these infrastructures more seriously, there would have been significant evidence of ongoing serious efforts to address their vulnerabilities with stop-gap systems or tactics, techniques, and procedures (TTPs).

FINDING AND RECOMMENDATION

Finding 3: Organizations that anticipate and respond effectively to potential capability surprises—such as the Navy's SSBN[9] Security program, the Air Vehicle Survivability Evaluation program (Air Force red team), and the Aegis Ballistic Missile Defense program (in response to Operation Burnt Frost)—appear to possess the following characteristics: senior leadership support; team independence; access to a strong base of cross-disciplinary technical and operational expertise; an ability to identify threats through campaign-level modeling, system-of-systems simulation, and high-fidelity physics-based models; precise vulnerability modeling and analysis capabilities validated by test and experiment data; mechanisms to recommend and/or deploy solutions as necessary; adequate, steady funding; and focus on a particular mission such as Navy SSBN Security. In addition, these organizations appear to leverage modeling, simulation, and analysis tools in conjunction with a network of experts to expose bias, offer critical review, model and test against potential vulnerabilities, and demonstrate alternative solutions to respond to surprise.

At the same time, assessments of threats to the critical technologies that enable U.S. naval forces appear to be conducted on a small scale rather than being quantified, modeled, and characterized for U.S. naval forces as a whole. For example, threats to precision navigation and timing sources or cyberattacks embedded within Navy weapon systems could impact a wide array of naval operations. However, U.S. naval forces as a whole do not seem to be utilizing the best methodologies for assessing surprise. One of these methodologies would be the creation of red teams, that could simulate or represent adversarial thinking across global cultures.

Recommendation 3: As its first tasking from the Chief of Naval Operations (CNO), the surprise mitigation office (see Recommendation 1) should (1) identify and prioritize any broad response to operational and technology threats that are not owned by any one mission authority and (2) establish threat study groups to characterize, quantify, and model these specific threats as well as leverage existing resources (modeling, simulation, and analysis tools and test data used by a network of subject matter experts in academia, industry, laboratories, and the Service colleges). The output from (1) and (2) should be disseminated to U.S. naval leadership as soon as possible. Careful attention should be paid to surprises not addressed by any program, or where a substantial gap exists between programs.

The CNO, the Commandant of the Marine Corps (CMC), and the Commandant of the Coast Guard (CCG) should take steps to ensure that red teams—with sufficient independence and expertise but, at the same time,

[9]SSBN, nuclear-powered ballistic missile submarine.

a fresh influx of participants—are able to depart radically from traditional thinking in order to help U.S. naval forces as a whole prepare for combat and develop new tactics. In particular, efforts should be made to expand and periodically refresh the composition of red teams to achieve a greater diversity in thinking and better represent the adversary.

4

Prioritization, Option Development, and Decision Formulation

INTRODUCTION

In order to develop cost-effective solutions to often complex, time-critical, unanticipated problems, there needs to be a process by which to generate mitigation options and pass them through a series of screens that reduce the many possibilities to a very few solid and attractive options with credible effectiveness and cost and favorable technical and operational viability metrics. This chapter outlines such a process within the proposed surprise mitigation office. The general process is outlined in Figure 4-1.

CONCEPTUALIZING RAW OPTIONS TO MITIGATE HIGH-RISK SURPRISES

The first step in solving a problem is to conceptualize as many solutions as possible that could be brought to bear on the problem. This spectrum of options should encompass materiel and nonmateriel solutions within the naval forces as well as solutions that could be provided by joint action with other services.

Understanding the Surprise Challenge and Its Potential Impact

Conceptualizing a possible solution must start with fully understanding the problem. At a minimum the conceptualization process needs to ask and answer, "Why would this happen?" "What vulnerability do I have that my adversary is exploiting?" "What are the key actions that he is taking to bring this about?" "What are the consequences of his actions on my ability to operate?" Once this understanding is established, a set of key metrics can be created that capture the

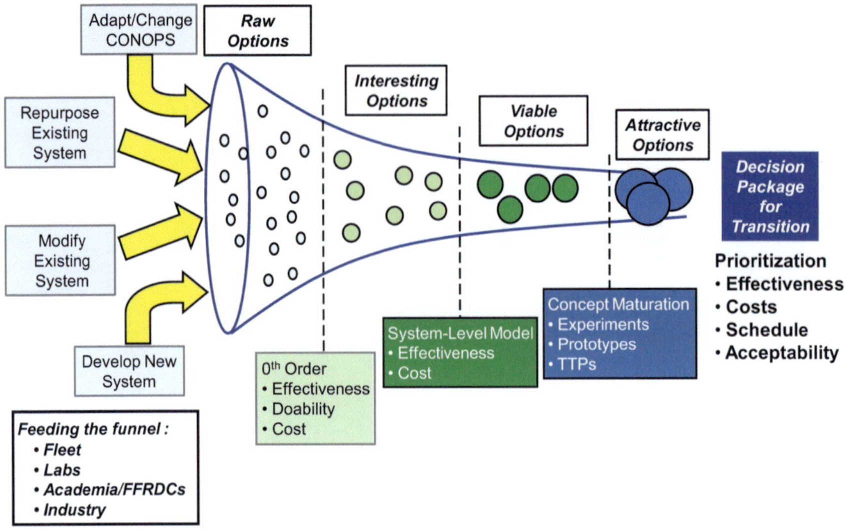

FIGURE 4-1 Creating, evaluating, and maturing options to address a high-risk surprise. CONOPS, concept of operations; FFRDC, Federally Funded Research and Development Center; TTPs, tactics, techniques, and procedures.

essence of both the adversary's ability to create the surprise as well as our own operational effectiveness in the presence of it. Both sets of metrics are important to establish early because they will form the foundation for measuring the operational effectiveness of each potential solution. The level of detail in both the questions and the metrics will grow as candidate options pass through the screening process shown in the figure and will become more mature, but they must have been thought about and must exist at some level, even at the outset.

The Conceptualization of Possible Options to Deal with a Given Surprise

Based on the investigation above regarding why there are potential consequences to what the adversary is doing, what those consequences are, what it is exploiting, and the actions it is taking to carry out its exploitation (and, inherent in that, what its exploitation vulnerabilities are), options can be conceptualized at a high level to deal with the adversary surprise. There are at least two dimensions to the types of solutions that could be brought to bear on the problem. One dimension is what can be done to eliminate or mitigate the problem. It includes options to

- Avoid the adversary's ability to exploit our existing vulnerability.
- Interfere with the adversary's ability to exploit our existing vulnerability by interfering with elements of the adversary's required effects chain, probably with some offensive action.
- Mitigate or nullify the consequences of its actions by reducing or eliminating our existing vulnerability.

Any of these three actions might involve actions by naval forces alone or in joint operations with other Service forces.

The second dimension involves how things can be done to address any of the three options listed above. It includes the following options:

- Adapting or changing TTPs without the necessity for any materiel changes.
- Repurposing existing systems to suit a new mission without any significant system modifications but perhaps some minor software modifications (at a scale similar to what was done for Burnt Frost).
- Modify an existing system to modify an existing capability or to create a new one. This could entail the use of existing or new science or technology.
- Develop a new system(s) to create a new capability that is not available through modification of any existing system. As in the above case, this could entail the use of either existing or new science and technology.

As solution options are conceptualized across both dimensions, it is incumbent on whoever is driving this conceptualization process to force the process to think broadly. Trying to create options at many of the intersections of the two dimensions above will certainly help, but in addition there is the notion of "hats" or "flavors" that can be brought to bear. A hat or flavor is a particular way of looking for solutions. Suppose that one is looking for an option with the lowest near-term cost. One would focus on solutions that cost the least during R&D or those that are nonmateriel solutions at the expense of other attributes. Suppose that one is looking for the most capable solution. It is unlikely (although not impossible) that it will also be the least expensive or the quickest to initial operating capability (IOC), but one could force an option to be very high performance at the expense of cost and time. Minimum complexity, minimum ship fit issues, minimum schedule, and minimum risk are exemplars of other flavors that could be used to force the initial conceptualization to be broader than it might otherwise be. Through the combination of specifically addressing the what's and how's and then thinking of flavors, a rapid and broad examination of options to avoid, eliminate, or mitigate the effects of a serious surprise can be accomplished.

Establishing the Value Proposition for Assessing "Goodness"

Inevitably the options developed above will differ significantly in terms of their important attributes. Some will be lower cost, some will be more effective at either eliminating or mitigating the impact of a given surprise, some will be able to be fielded far more rapidly, some will more readily fit the culture or existing ways of conducting operations, and some will be more mature and represent lower risk, etc. Equally important to the differences in the attributes themselves is the fact that depending on the nature of a given surprise, the relative importance of each attribute will vary. In screening out options that are less desirable than others and ensuring that the better ones survive for further scrutiny and development, it is critical that a value structure be defined that, for a given adversary surprise, captures those solution attributes that are important and assigns a relative weighting to each. This weighting can be qualitative (e.g., critical, very important, less important, not important), but it needs to be established to deal with the fact that for a given problem, all attributes are not created equal. As for the attributes themselves, they are generally easy to define, and at a minimum should include such key attributes as these:

- Effectiveness against the impact of the surprise being countered.
- Robustness to variations in the particulars of the surprise.
- Reliability—particularly in the initial screening this will not be given a quantitative treatment, but there may be some characteristics of an option that make it inherently more or less reliable than other options.
- All elements of cost—development, acquisition (if any), support, and the like. Note that there may be different weightings even within this element, because near-term resources are often far more precious than those in out-years.
- Time to field or procure an initial capability to counter the surprise.
- Risk in achieving this capability—for example, Are there unknowns at this time? Is this dependent on technologies that at this point are immature? Are there ship fit issues that need to be resolved for which there are not ready solutions? Are there uncertainties in the overall impact on the Navy of a recommended change in CONOPS.
- Adaptability and flexibility in assessing a variety of options.
- Other—What is in a particular attribute that will differ from one surprise to another? (There will always be some issue that pertains to another.)

CANDIDATE OPTION EVALUATION

There needs to be a structured mechanism to screen the options down to a workable few. By "structured," however, the committee does not mean that judgment needs to be replaced by a quantitative treatment of attributes. Rather, human judgment pervades the process and should not be replaced by "bean counting." The structure to be imposed is one that captures the essence of the screening process, retains the important why's that screen certain options out, and sends others on to more refined analyses, but that does not interfere with the use of good judgment and time-tested experience throughout. This process is described in two steps below.

Initial "0th Order" Evaluation (Many Options to Consider: Weeding Out Those That Are Not Interesting)

This level of evaluation needs to be accomplished at a very high level, based largely on a few subjective or at best semiquantitative measures. The committee focuses on high level and subjectivity because at this stage there are likely to be many options to consider, there will not be a lot of detail defining each one, and there will be many unanswered questions associated with each, and rapidly and inexpensively screening out those that will never be attractive regardless of how much detail is provided is the goal. Experience shows that it is always exceedingly difficult to do this and is critically dependent on the judgment of the people involved. Notionally one would set as an objective in this phase to get the surviving options down to a handful or so, because the more detailed examination and maturation activities that follow this phase cannot afford, either in time or resources, to deal with more than a very few. A secondary goal in this phase is to provide the initial insights and articulate the questions to be answered on those options that survive this initial screen.

Nevertheless, however high level and however subjective the criteria for this initial screening, they need to track in some fashion the attributes described above, perhaps focusing at this stage on effectiveness, cost, time, and feasibility. In order to effectively use high-level measures to screen out less attractive options and to resist the ever-present temptation to take just one more look, the recommended surprise mitigation office needs to be populated with at least a few very good and very experienced operations analysis personnel who understand the Navy, understand the systems and technology, think at a systems level, and are comfortable with back-of-the-envelope analysis and judgment when it is called for and more detailed modeling and simulation (M&S) when it is appropriate. One of the biggest challenges in the creation of the office may be to find, attract, maintain, and refresh a critical mass (perhaps a half dozen) such people.

More Detailed Screening: Assume Five Options That Are Interesting)

The few options that survive the initial screening need to be subjected to lower level and more rigorous analyses based on system level cost and effectiveness modeling and the initial treatment of some of the other considerations highlighted above that were not treated in the first-pass screening. Sufficient detail needs to be established for each of the options to enable meaningful modeling of the interaction between the victim of the surprise, the effects chain of the "surpriser," and whatever solution is being investigated. This needs to be accomplished with sufficient system fidelity to distinguish one option from another and at a sufficient level to highlight broad capability differences and effectiveness. The same is true of the cost modeling in this phase—it must be specific enough to highlight differences and detailed enough to be believable, but not so detailed that it requires inputs that are far beyond what is appropriate at this point of definition. First-order differences should be sufficient and should guide the level of detail asked for and provided. Providing the inputs for the cost analysis will, by definition, allow a first cut based on schedule requirements and an estimate of how long it will take to get an initial capability into the field. Lastly, all of the other important considerations in the value structure established for this problem should also be considered for this second cut. The objective should be to filter down to a smaller number (say, three) candidates.

CONCEPT REFINEMENT AND PROOF OF PRINCIPLE (ASSUME THREE VIABLE OPTIONS)

The two things accomplished during this phase are described below: One deals with residual unknowns in the key enabling elements of each approach and the other deals with operations related to each approach.

Experiments and/or Prototyping of Key Elements of Each Option

It is likely that a few significant questions may still be unanswered at this point: Can a certain key performance characteristic be achieved? Can some key physical attribute be met? Is the assumed production cost for a high-volume component feasible? Can any one of a variety of other key components be counted on to materialize with a reasonable amount of development effort? Other questions related to field operation of a system may also remain unanswered, such as, Can an element be set up and become operational at the desired performance level within some stressing time line? Will it interfere with (or be interfered with by) other systems in close proximity on the same platform? Is the workload for an operator such that the assumed number of operators in the predicted operations and maintenance (O&M) cost is unrealistic? Still other questions may remain about the interoperability of one of the options with some other system outside the control of the system designer. These and many other potential critical questions can be answered by performing a key experiment or by prototyping a key element or component of the system (by, for example, using three-dimensional printers for parts on demand). If the viability or desirability of a particular concept hangs on one or two issues about which some doubt or ambiguity remains, then the objective of this task is to remove this doubt at far lower cost and in far less time than simply waiting to find out later in the development cycle. The key requirement is to design the experiment or the building of the prototype to maximize learning on the important unknowns rather than as a show-and-tell to hype the best features of the system.

Rethinking the TTPs Enabled or Required by a Given Approach

Given that the system options at this point in the process are down to a small number, it is worth taking a further look at the TTPs for each option. If an option calls for a fundamental change in the standard TTPs associated with a mission, then the impact of that change on doctrine, training, or the conduct of other related missions needs to be considered. There may also be something associated with a system concept that, while not required, would enable a change in TTPs that provides additional warfighting leverage. In this case the added benefit needs to be noted and credit taken for it. Regardless of whether the impact of the system on TTPs is constraining or leveraging, it needs to be understood to avoid major warfighting "surprises" down the road should a given system concept be pursued further. It is this interaction between the technical and the operational, and between the engineers and operators and warfighters, that is as necessary to the selection process as is the technical refinement of a potential materiel solution.

PRIORITIZATION: THREE OPTIONS— WHICH IS THE MOST ATTRACTIVE?

If a more detailed understanding of the key attributes and characteristics of each option has been obtained, no showstoppers have been identified, and all three options have survived, the next question to be answered is, How do the three options compare? What is their rank order of "goodness"? Which one (or in rare circumstances, ones) should be developed? At this point, three steps need to be followed.

Establish Relevant Metrics for Each Option

Earlier in the process a number of high-level measures of effectiveness (MOEs) were established to evaluate and cull the initial concepts that were posited as solutions. These MOEs were at first rough estimates of performance and effectiveness, cost, schedule, and risk. Through the refinement process, more detail is available on each of these measures, particularly on those measures that distinguish one option from another. Further, let us assume that the development program that has been laid out for each option captures the inherent risk of that option in terms of the program's content, so that the nonrecurring development cost and schedule drives the risk down to an acceptable level. Thus, differences in development risk have been translated to differences in development cost and schedule and do not have to be treated explicitly. What therefore may remain as critical metrics are the following:

- Performance, effectiveness, and robustness;
- Development, acquisition, and life-cycle cost;
- Time-to-field capability;
- "Impedance" or cultural match to naval forces; and
- Other considerations that are deemed important and are not captured in any of the above.

Based on the level of detail available for each option, these measures need to be defined one or two levels below this level and should form the basis for evaluation. Using a set of M&S tools appropriate to the level of detail available, each of the options would then be evaluated based on the metrics listed. As lessons are learned through the evaluation process, such as why a particular option performs less well than another or has a much longer development cycle or is more counterculture to Navy standard practices, they need to be captured and documented, both so that the reasons behind a particular metric are not lost but, even more important, so the process of ranking the options becomes more transparent.

Establish Relative Attractiveness Based on the Value Proposition or Structure

Once the various metrics have been established to distinguish the various options, it is often tempting to automate the ranking of the options based on some simple arithmetic algorithm combining the metrics and weightings. This temptation needs to be resisted, because human judgment is as much a part of the process as is the quantitative evaluation of MOEs. Rather, using the quantitative measures as guides rather than absolutes, the evaluators should stand back and try to see what high-level messages, if any, are present within the individual assessments. Which highly ranked options best satisfy the criteria on the value structure? Which one or ones have deep holes in some of the measures? Are those holes critical and not easily filled? What about a concept that ranks very high on all of the most important measures but falls short on a few of the others? Is there a work-around for them? It is this kind of questioning, answering, and then re-synthesizing that should be spurred by looking at the quantitative MOEs to come up with the best informed basis for the final recommendation.

DEVELOP TRANSITION DECISION PACKAGE

The final step in the synthesis and evaluation process is to develop a draft briefing package and written report containing at a minimum the background on the problem being addressed, the alternative solutions that were examined, the evaluation process to which they were subjected, the value structure that was used to weigh the various criteria or MOEs, the rank ordering of the alternatives that resulted, and the recommendation for transitioning the concept for further development, testing, prototyping, or fielding, whichever is appropriate. It is also worth noting that the most important reasons are captured in readily available backup should questions arise on any of the important issues covered in the briefing. These same reasons should be part of the discussions in the report.

The team should engage with the organizations that are key to the success and acceptance of the recommended solution or have a major interest in it. Based on the give and take at those discussions, the briefing and report should be updated as necessary to accommodate whatever feedback was obtained and was determined to be valid and useful in terms of the final product.

The final step in the process is to develop an executive-level decision package for whichever office is the final authority for determining whether to proceed with the recommendation. The content and level of detail should be appropriate to the office although the options considered, how they ranked, and the key rationale for the recommendation should all be treated.

For there to be an effective solution, the enterprise as a whole will have to respond in kind. It is important to include the Navy contractors involved, and their buy-in will be important for success. This is a tough problem. Traditionally the Navy has relied on its laboratory community—naval centers, university-affiliated research centers (UARCs), and FFRDCs—for objective advice (advice without a profit motive). Also, multiple labs are typically involved in removing the possible systematic technical biases of any single expert.

FINDING AND RECOMMENDATION

Finding 4: With the recent establishment of the Strategic Capabilities Office within the Office of the Under Secretary of Defense for Acquisition, Technology, and Logistics (USD AT&L), there appears to be an opportunity for a surprise mitigation office to provide naval force component solutions to surprises facing the entire Joint Force. Furthermore, there is an opportunity for both offices to draw on each other by sharing expertise, methodologies (modeling, simulation, analysis, red teaming), and learning.

Recommendation 4: The Chief of Naval Operations (CNO) and the Under Secretary of Defense for Acquisition, Technology, and Logistics (USD AT&L) should encourage their respective "surprise offices" to develop and foster a close working relationship with each other. In particular, the CNO and USD AT&L should direct their surprise offices to share ideas and to collaborate on methodologies (modeling, simulation, analysis, test data, and red teaming) in the interest of efficiency and obtaining consistent and coordinated results. Technical interchange meetings and the frequent exchange of information between the two offices and others that may be eventually established by the other Services should also be encouraged.

5

Resource and Transition Planning

This chapter will address how strategic naval planning, budgeting, and policy impact the prioritization of risks and the mitigation options identified in the previous chapters, as well as budgeting and resourcing for a response. Several broad observations are offered on the limitations of the current programming, planning, budgeting, and execution (PPBE) system to ensure that naval forces are resilient, flexible, and responsive to capability surprise or, conversely, able to deliver it. These observations are followed by considerations for modifying policy to address these limitations. The committee urges a naval-forces-wide awareness of the following common negative responses to surprise in the e-commerce industry:

- First, denial that an event occurred;
- Second, a misunderstanding of the event's effects; and
- Finally, embarrassment that the organization is surprised but not admitting to any mistakes.

An awareness of these responses is an important step to overcoming them.

THE PPBE SYSTEM AND SURPRISE

The PPBE system is a requirements-based, rational decision-making process to determine cost-effective solutions in force structure that are sufficient to satisfy future defense-related scenarios and degradation of capability. The system's characteristics, however, usually inhibit it from imparting enough resilience, flexibility, and responsiveness to allow the naval forces to respond to disruptive or tactical surprise. A requirements-based decision process identifies gaps or needed capabilities in response to a known or projected need but will fail to invest in response to unknown, unanticipated surprise or surprises that only partially fall within the system's mission. Even if aware of new threats that could emerge, the PPBE system might label them as "acceptable risk" owing to fiscal constraints. These acceptable risk items may later emerge as an intelligence-inferred surprise with much more serious effect than had been predicted.

Every year the PPBE system prepares scenarios modified only slightly to reflect current defense strategies and intelligence on new threats.[1] This scenarios framework is susceptible to escalation of commitment, which can lead to continued investment in expensive technologies whose capability may have degraded or to potentially faulty courses of action. A frequent cause of escalation of commitment is making marginal trade-offs between platform capabilities, which may be cost-effective within a set of defense planning guidance scenarios and their concept of operations but creates a brittle force for unanticipated challenges. A historical example is the removal of guns from fighter aircraft in the early 1960s because it was assumed that missiles would reign supreme. Then, guns had to be reinstalled on fighter aircraft to conduct close-in dogfighting during the Vietnam War, when the missile technology proved inadequate and the concept of how air-to-air battles would play out proved faulty. A recent example is removing long-range surface-to-surface missiles from the Navy's surface ship weapon inventories. This may be a rational decision in light of funding constraints and air power's availability to cover long-range sea strikes in known scenarios, but it also makes the surface fleet less capable of independent offensive action if an unforeseen surface-only challenge or opportunity emerges.

The peacetime budgeting process allows no "wedge" of undesignated funds for rapid response. During both the Vietnam War and the recent Afghan and Iraqi campaigns, DOD benefitted from congressional supplemental budgets—some with the flexibility for secretary-level discretion on how to obligate funds in response to unanticipated events. During peacetime, however, these supplements do not exist and every last budget dollar is assigned to a specific line program. Therefore, any in-year response to a disruptive or tactical surprise must be funded from existing programs, creating an extra decision process on where to cut resources. Then, funds are identified to address a surprise, no entity is empowered to oversee an overall response to capability surprise.

MITIGATING RISK OF SURPRISE WITHIN THE PPBE PROCESS

A wholesale revamping of the PPBE system to address surprise mitigation is probably not possible. However, scanning discoveries and assessment and mitiga-

[1]LTC Boyd Bankston and LTC Todd Key, USA. 2006. *White Paper on Capabilities Based Planning*, Draft, Presented at the MORS Conference, Capabilities-Based Planning II: Identifying, Classifying and Measuring Risk in a Post 9-11 World, McLean, Va., March 30. Available at http://www.mors.org/userfiles/file/meetings/06cbpii/bankston_key.pdf. Accessed May 30, 2013.

tion alternative analysis should be integrated into the PPBE cycle. One way of doing this is to have the red team's results during the last budget cycle integrated into the directions for the next cycle. Another is end-of-cycle reviews that explicitly consider ways to increase the resiliency of the naval forces to anticipated and unanticipated surprise.

As an illustration, consider what could happen if there is a denial of access to space: forces might operate in a degraded electromagnetic environment owing to the loss of satellite information. Seizing individual command initiative to meet the well-understood intent of the commander could be one good way to respond to this anticipated surprise. Individual command initiative and innovation during wartime should be ingrained in the naval forces culture through training, education, evaluation, and reward. However, individual command initiative must also be supported by technical capability at the outset by ensuring, as much as possible, that each command level is self-reliant in executing its own weapon systems' kill chains. For example, a cost-effective budgeting process would devalue individual ship or unit organic aerial intelligence, surveillance, and reconnaissance (ISR) as redundant and would favor shared resources to support search and detection among several ships or units. The shared resource, however, becomes a liability when it is lost because of communication failures or enemy fire. Likewise, a shared timing source needed to provide coherence between ships/units and weapon systems makes sense, but only until that timing source is lost. A local master timing clock for single and group ship/unit operations will add resiliency to locally controlled operations in a degraded electromagnetic environment. A dedicated step inside the naval forces PPBE cycle to increase force resiliency by enhancing an individual command's ability to be self-reliant is warranted. Local ability to conduct a full kill chain with organic assets is a good litmus test for enabling command initiative.

A POLICY OF RESILIENCE TO MITIGATE RISK AND ENHANCE RESPONSE

A considered policy to review weapon system rigidity resulting from decades of cost-effective planning, marginal trade-offs within set defense planning scenarios, and high-end platform design is warranted. There are three ways to build resilience into weapon systems that are subject to capability surprise:

- Enhance weapon system/platform design to include the capacity for quickly adding or modifying capability,[2]
- Increase the mix between high-cost multimission weapon systems/platforms and less expensive single-mission weapon systems/platforms, and

[2]Defense Science Board. 2011. *Report of the Defense Science Board 2010 Summer Study on Enhancing Adaptability of U.S. Military Forces*, Part A: Main Report, Office of the Under Secretary of Defense for Acquisition, Technology and Logistics, Washington, D.C., January.

- Build redundancy into the weapon systems/platforms to accomplish required missions.

In the face of budget cutbacks, the last option will not be explored, because increasing weapon systems/platform redundancy is the most expensive option.

Designing weapon systems/platforms so their technical capability can be rapidly added to or modified (without identifying that capability) will provide options to deliver or respond to a disruptive surprise. A successful example of excess capability build into platforms is the Spruance-class destroyer family. When commissioned, USS *Spruance* was about 8,000 tons and described by a quote attributed to VADM Mustin: "I can walk a country mile on that ship without finding a single weapon system." The last Spruance destroyer, the USS *Cushing,* when decommissioned displaced 9,800 tons, carried vertical-launched Tomahawk missiles, had a Kevlar-enforced superstructure, and possessed advanced active and passive antisubmarine warfare (ASW) systems not even envisioned during the original programmed buy. Extra space allows for needed capability growth, particularly in warships expected to stay in service for 25 or even 40 years.

Enhancing the capacity to add capability may also be extended to onboard weapon systems/platforms, sensors, and electronics. The Littoral Combat Ship class, although initially challenged with production issues, is exploring modularity for entire missions. If the platform and mission module integration issues are resolved, the committee views the flexibility to create new mission modules in response to unforeseen operational and technical challenges as an advantage for this program.

The issue of weapon system/fleet mix has led to continual debate among naval historians, academics, and professionals. In a cost-constrained environment, when envisioning a low-threat operating environment or scenarios where there is little risk of loss of life or ships being damaged, building fewer but more expensive weapon systems/multimission ships is economically rational. More capability can be provided at sea with fewer hulls, crew, logistics, and total life-cycle costs. When, however, there is a risk of that ship being damaged or sunk where there is a high probability of tactical surprise attack, the reverse is true—that is, cost-effectiveness becomes "too many eggs in one basket." A damaging hit on a guided missile destroyer (DDG) hull is degradation not just of the fleet's air and missile defense capacity but also of ASW; antisurface, maritime interdiction operations; support for Marines ashore; and helicopter-related-missions as well. Building more less-expensive, single-mission ships may increase fleet resilience, to absorb the impact of an unanticipated threat at sea, and provides more options for response through geographic dispersion as well as greater ship availability for quick modifications. The committee does not endorse a complete overhaul of fleet composition, but it encourages a review of the force mix with the added considerations of the surface fleet's capability apart from high-value unit protection and ballistic missile defense (BMD), and resiliency to surprise as the determining factor.

OPERATIONAL SCENARIOS WITH ANTICIPATED SURPRISES AND POLICY IMPLICATIONS

When the three surprise scenarios were selected for inclusion in this report, there was concern about several operations where issues were known to exist that could provide the catalyst for future anticipated surprise. Historical review and wargaming these scenarios might raise further concerns that should be addressed by aggressive policy action to mitigate the known issues. These scenarios include the following:

- Humanitarian assistance and disaster relief (HA/DR) or a major U.S. government policy change,
- Global maritime security operations, and
- Maritime defense of the United States.

The issues related to contingency response by HA/DR primarily involve defining roles and coordinating efforts of a number of naval forces. For example, establishing formal lines of communication and coordination domestically between U.S. Coast Guard (USCG) districts and Navy regions and conducting exercises to develop contingency plans will foster a better command-and-control system to use during a domestic natural disaster. Internationally, USCG vessels are accepted much more readily than U.S. Navy ships by governments requiring assistance. For example, USCG ships were quickly granted entry into the Black Sea to assist the Georgians following the conflict between Russian and Georgian forces in the former Soviet republic. Fleet commanders should be encouraged to review their areas of responsibilities for regions where this is concern and to plan for USCG inclusion into the response. The responsiveness and flexibility naval forces have demonstrated in the past is recognized, such as using aircraft carriers as Army helicopter platforms in Haiti. In fact, the inherent excess capacity resident on an aircraft carrier without its air wing allowed for its use as an Army seabase and reinforces the argument in the previous section that excess capacity creates flexibility in response to unforeseen events.

An unanticipated major U.S. government policy change—one that deviates from long-established contingency or war plans and requires new or revised plans that have not been tested, wargamed, or practiced—was also raised as a possible surprise on naval forces. Although the fleet is well equipped and experienced in crisis action planning for operations, the operational commands do not have the ability to assess the implications of major strategic shifts in a timely manner. Here, naval educational assets like the Naval War College and Naval Postgraduate School could be tapped to lead policy change and fleet assessments using wargaming and campaign analysis.

The other two scenarios, where threats are known yet organizational response issues exist that require policy attention to avoid intelligence-inferred surprise, are related to maritime security, both globally and in defense of the homeland.

Globally, the United States is economically dependent on the free flow of goods and communications on and under the sea. Closure of a major shipping choke point or severance of an undersea cable terminal should be assessed for its impact on the economy and potential mitigating responses developed; at the same time, possible courses of action to deter the event from happening should be taken. A response will require coordination between the United States and regional governments, which will be accelerated if fleet commanders consider such events now and establish contacts within regional navies and coast guards of concern. Our nation's ports and waterways at home are vast, challenging the USCG to provide sufficient federal coverage. Although the Coast Guard works closely with state and local law enforcement agencies on patrols and response, a major security event in a waterway or port might require the USCG to request Navy or Marine Corps assistance in addition to their support for the U.S. Northern Command through the Naval Component Commander. This is not a familiar operation for these forces and should be considered, gamed, and practiced.

ORGANIZATIONAL AND BUDGET IMPLICATIONS

As recognized in other areas of this report, capability surprise impacts missions, while the PPBE and acquisition programs are weapon-system- and platform-centric. Likewise, the combatant commanders are concerned about mission accomplishment, not platform delivery. This creates an inherent hurdle for the budget process to respond to a capability surprise. Because our defense budget is public knowledge, its constraints provide a transparent window onto our strategy formulation, which hostile countries may use to develop their own asymmetric naval strategies.

To overcome these shortcomings in the face of preparation for surprise, the naval forces need to define and develop a resource authority responsible for the prioritizing, decision processing, and resourcing the responses to capability surprise. This authority should be informed by the scanning, assessment, and mitigation analysis efforts. It should be concerned with force and fleet resilience to meet unanticipated surprise, communicate frequently with commanders, and have direct influence on the budgeting process. In addition it should oversee rapid prototyping, tactical development, and acquisition program adjustments and should be in a position to force the introduction of any capability to respond to or deliver a surprise. This authority must have influence to coordinate a naval-wide solution (USN, USMC, and USCG) and coherence among service platforms. It should have the benefit of funding with discretionary power to allocate quickly for rapid response. Outside of the past customary supplemental funding from Congress, this ability does not exist. Therefore, another role of the authority will be to determine the best options for ongoing funding of these activities. This would be the role of the committee's recommended surprise mitigation office.

As an illustration for the surprise mitigation office's role, the committee uses Scenario 2—a social media crowd emergence. If scanning, assessment, and mitigation alternative analysis efforts show that an adversary has the ability to use social media to inspire a crowd to emerge in a location that would impede or embarrass the United States, the authority would review the current tactics, technologies, and procedures (TTPs); query intelligence-monitoring capabilities in social media; assess with the commanders the potential impact; and evaluate current and programmed programs to intrude, misdirect, and/or respond in the social media. If there is a funding shortfall, this threat would require prioritization against other potential vulnerabilities for resources. Once resources are assigned, the authority would oversee any technical prototyping, testing, and force introduction.

INTEGRATION AND INTEROPERABILITY

The committee learned that the Vice Chief of Naval Operations (VCNO) is considering creation of interorganizational teams to set warfighting capability baselines (WCBs) in order to address system-of-systems integration and interoperability (I2) to enhance mission effects. This effort would be under the guidance of the Office of the Chief of Naval Operations (OPNAV) N9I and N2/N6. The committee applauds the proposal as an effective intermediate step toward creating the authority called for in the preceding section. An annual review in the form of WCBs of naval-forces-wide missions and prior to the program objective memorandum (POM) process allows a venue for an existing knowledgeable team to address short-term capability surprise through a range of responses, from change of tactics to resource allocation. Use of the systems commands' technical authority to enforce interoperability requirements for platform program managers should be combined with an award structure for the program managers to inspire a more systems-oriented approach and mitigate self-imposed surprise—namely, mission degradation due to interoperability shortcomings. The I2 proposal should be a top priority for Navy leadership to immediately enhance warfighting at modest cost. Once implemented, I2 should be shared with USMC and USCG leadership for possible integration across the naval force structure.

PRESENCE VERSUS PREPARATION

A consistent theme is that the demand on the fleet's time to conduct maintenance, complete basic individual training, and finish workups to meet deployment requirements leaves little time for systems and tactics training, red teaming, true experimentation, or time to think about reaction to unanticipated events. With a decreasing budget and fewer ships, this problem needs considerable attention. If the desire to maintain a high level of forward presence with a smaller fleet affects the preparation for a complex naval conflict, then how will new tactical and operational challenges be uncovered and addressed? A Center for Naval Analyses

(CNA) review of historical U.S. naval deployment strategies may provide alternatives to the current deployment pattern, which has remained basically unchanged since the Second World War.[3] With a smaller fleet and fewer sailors, deployment cycles may have to be rebalanced, with more time for true preparation.

RESOURCING IMPLICATIONS

This section of the chapter reviews the current status of resourcing for preparation and response, including the difficulties of explicitly resourcing programs to prepare for, counter, and create surprise. Preparing naval forces to deal with capability surprise is implicit in many of the routine resourcing activities that shape the standing force.

The resourcing process is highly structured, at least in principle, to field a force capable of dealing with a range of missions or contingencies that flow from the National Security Strategy, as shaped by statutory mandates and service doctrine. These directed missions and contingencies define the capability envelope within which the adequacy of our naval forces' capabilities is routinely assessed. However, the boundaries of this nonsurprise envelope are sufficiently imprecise and the assessment tools rudimentary and incomplete such that there is always some doubt that the risks of accomplishing the intended tasks are acceptably low.

Said another way, the pressures to further improve directed capabilities create a demand for such capabilities that usually exceeds available resources, leaving little room for the resourcing of preparations for surprise. But some such resourcing does exist. For example, the specifications for some types of military equipment include requirements for operations in environments and situations that are more demanding than found in the routine mainstream planning scenarios. Examples include provisions for operability in arctic conditions and in situations where nuclear weapons have been used or environments that contain chemical and biological hazards.

At the other end of the spectrum, some farsighted operational commanders do indeed use discretionary funding to conduct outside-the-box experiments and training exercises that provide a modicum of hedging against surprise. And some fraction of service science and technology funds is routinely allocated to investigating nontraditional approaches to potential but ill-defined future needs.

Overall, however, the resourcing of preparations for dealing with surprise, both self-imposed and external, is minimal. In the absence of explicit guidance, such preparations as may exist and be proposed are constantly threatened by mainstream demands. The result is a national maritime posture that is at excessive risk of being unprepared to deal with the types of surprises that have arisen in the

[3]Peter M. Swartz with Karin Duggan. 2011. *U.S. Navy Capstone Strategies and Concepts (1970-2010): A Brief Summary*, Center for Naval Analyses, Alexandria, Va., December.

past and will surely arise in the future or that is incapable of creating surprises that can be delivered to an opponent.

Members of the committee are well aware that design and payload margins are expected of mission platforms as part of the acquisition system. For example, even decades ago computing hardware was specified with a significant reserve, and ship centers of gravity and mast clearances were designed in anticipation of likely additional payloads. Further, as stated in the text, classes of ships such as Spruance and littoral combat ships anticipated modular changes as part of the design philosophy. Still, in recent years the need has arisen for platform life extensions that require additions never anticipated when the platforms were designed. For example, addition of the missile defense mission to the Burke-class Aegis destroyers is driving a need for radar aperture larger than can be accommodated. Further, the power and cooling required of a possible future addition for directed-energy weapons may be accommodated on the new DDG-1000 class but would be constrained on a Burke class. It therefore appears to the committee that the Navy should undertake a more comprehensive look at potential margin needs for the much longer planned lives of platforms. This is discussed further in Chapter 6.

FINDING AND RECOMMENDATION

Finding 5a: Management processes by which resourcing decisions are made and that potentially impact preparations for capability surprise among U.S. naval forces appear to be inadequate and to lack reserve capacity. In particular, there appears to be limited flexibility in the way of design margins for platforms and payloads to respond to a range of potential capability surprises, and it further appears that the Department of the Navy's investment in science and technology is insufficient to provide a robust array of technology building blocks that allow a rapid response to a broad range of potential surprises.

Recommendation 5a: The Chief of Naval Operations (CNO), the Commandant of the Marine Corps (CMC), and the Commandant of the Coast Guard (CCG) should add to their respective program planning guidance an explicit provision(s) that allows them to resource unit considerations and equipment design specifications, such as adding some adequate design margins into platforms and payloads in response to potential capability surprise.

6

Implementation and Fielding

INTRODUCTION

This chapter discusses the challenges associated with implementing and fielding a solution to a capability surprise. Implementation and fielding begin with a program plan and end with the deployment of a new capability. The importance of flexibility, timeliness, and affordability to capability surprise and how the existing acquisition structure can support those needs are discussed. The concept of open architecture is reviewed and how it is important to implementing capability surprise solutions through the concepts of repurposing and spiraling in new capabilities.

Needs

Surprise is difficult to predict, as discussed previously in this report. When it does materialize, the ability of naval forces to react effectively is dependent on three important principles: flexibility, timeliness, and affordability.

Flexibility

Flexibility deals with the ability to redirect and manage existing resources effectively in the face of surprise. Existing processes for acquisition afford us the flexibility to respond effectively, but we fail to take on the challenges of using this built-in flexibility because we are risk averse. The design and development processes have their waiver procedures, but many times programs prefer to manage to 100 percent of the requirements rather than a “good enough” solution that is more timely.

Rapid acquisition procedures have been used effectively during the wars in Afghanistan and Iraq. However, based on these experiences, the logistics and support services need improvement and must be adequately addressed in future conflicts.

Finally, more flexibility must be incorporated into the budgeting process to allow for capability surprise. Restrictive budget planning and allocation does not allow for the resources to address unexpected surprises. The development, test, and acquisition communities need to have more flexibility to allocate reserves and/or reallocate existing funding without the delays inherent in the existing programming, planning, budgeting, and execution (PPBE) process.

Timeliness

Addressing the capability surprise challenge is very similar to addressing the needs that have created the Joint Urgent Operational Needs Statement (JUONS) process. In both instances one is challenged to provide the operational warfighter with a capability that is lacking in the face of an unexpected adversarial threat and to answer that threat in as short a time period as possible. The JUONS process generally entails looking for a solution to a known enemy capability for which we do not have a response. It is real, immediate, and usually significantly impairs the warfighter's ability to freely operate. The capability surprise challenge can be categorized into three different elements based on the time horizon of the threat, defined as follows:

- *Urgent.* 0-2 years response horizon.
- *Emergent.* 0-5 years response horizon.
- *Deliberate.* 2-6+ years response horizon.

When it comes to urgent surprises, hostilities are most likely already under way, and solutions to unanticipated threats from our adversary are needed and being pursued. This is very similar to the scenario for the JUONS requests.

Emergent surprises are different from urgent surprises in that they are often proactive responses to estimated threats during peacetime conditions. There is assumed to be some time period in which one can prepare a response before one expects to have to address it under operational conditions. There is a limited time period one has to prepare the new response, test and train with it, and then deploy it in anticipation of the enemy's threat. In times of active conflict, efforts to prepare for emergent surprises will merge with efforts to prepare for urgent surprises, especially for early-stage initiatives. In this type of scenario one could find oneself both preparing new capabilities to rapidly field against observed surprises (urgent) as well as proactively pushing new capability to the field in anticipation of estimated new capabilities of the enemy (early-stage emergent).

Affordability

Figure 6-1 indicates how the number of acquisition professionals has declined over the last 20 or so years while procurement dollars have increased over the same period, primarily owing to the ongoing wars. Given this trend, a key to improving the rapid acquisition of solutions is the quality and type of the staff in these positions. Simply slashing a workforce already overloaded with demands makes it difficult to apply the innovative thinking necessary to address the acquisition needs for capability surprise. If the staff are focused on work flow, they will become very process driven, impeding the innovative thinking needed for fielding a rapid solution. This will breed bureaucracy, where the letter of the requirement or contract will become the driving factor rather than the time to fielding. A properly balanced workforce is required to ensure that innovative thinking is brought to bear and will provide managed risk solutions in a timely manner to our warfighters.

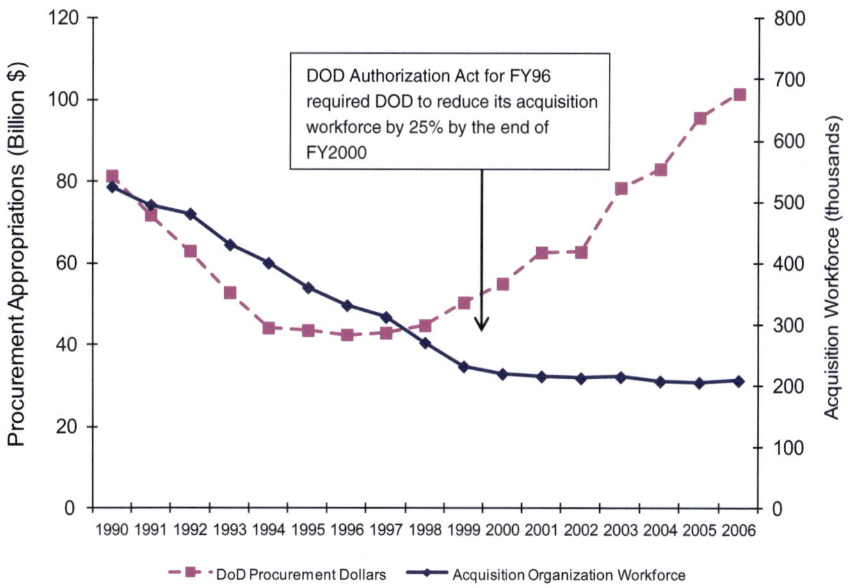

FIGURE 6-1 DOD acquisition workforce: SOURCE: Jacques S. Gansler, University of Maryland, "Fulfilling Urgent Operational Needs," presentation to the committee, Irvine, Calif., June 27, 2012. Source of workforce data: DOD IG Report D-2000-088, February 29, 2000, and DOD IG Report D-2006-073, April 17, 2006. Source of budget data: Annual Defense Reports, available at http://www.dod.mil/execsec/adr_intro.html. Procurement supplementals for FY2005 and FY2006 not yet reflected in Annual Defense Reports were obtained from Congressional Research Service Reports (Defense Science Board, 2008).

Using the Acquisition Process to Be Responsive to Capability Surprise

A natural reaction in response to delays in fielding new capabilities is to point at the DOD acquisition system and address changes through an update to the DOD 5000 procedures.[1] Traditionally this update has focused on the Federal Acquisition Regulation System/Defense Federal Acquisition Regulation Supplement (FARS/DFARS) procedures with a particular emphasis on the requirements oversight process—for example, the Joint Requirements Oversight Council (JROC) and the Joint Capabilities Integration and Development System (JCIDS). This committee takes a different view of the acquisition challenge in the face of capability surprise. It focuses less on the procurement process and more on the way we ask industry to develop and provide capability. The answers must not only be capable but must also be timely and affordable for the military and industry alike.

REPURPOSING

Repurposing Platforms—How Repurposing Has Worked in the Past

Repurposing in the naval forces and the military in general is not a new concept. It has been successfully applied in numerous instances and has saved the nation a fortune. It also has permitted rapid and timely redeployment of assets to meet new threats and resulted in incredible longevity for important platforms. In many cases, the repurposed "vehicles" were robust and large enough to accommodate payloads and purposes that were never foreseen or planned when they were first designed.

B-52 Stratofortress

The B-52 (Figure 6-2) was introduced in 1955 as a high-altitude nuclear bomber. It was repurposed during the Vietnam conflict to drop conventional bombs from a high altitude. It was again repurposed during the cold war as a low-altitude conventional bomber (while keeping its original mission as a nuclear bomber). During the 1980s, the B-52s had a stand-off mission when they were equipped with air-launch cruise missiles (ALCMs). During the cold war, they were repurposed to carry other weapons and to deploy mines. During the first night of Desert Storm, two B-52s flew the opening stages at 500 ft. In Afghanistan and Iraq, the B-52s were again repurposed to provide close air support by

[1]There have been many reports that have made recommendations to address systemic DOD acquisition issues. For example, see National Research Council, 2010, *Information Assurance for Network-Centric Naval Forces,* The National Academies Press, Washington, D.C.; National Research Council, 2004, *The Role of Experimentation in Building Future Naval Forces,* The National Academies Press, Washington, D.C.; and http://acquisition.navy.mil/home/policy_and_guidance.

FIGURE 6-2 B-52. SOURCE: U.S. Air Force.

the addition of targeting pods and smart weapons. These remarkable 57-yr-old platforms are expected to remain in service for another 15-30 years, giving us an effective platform for nearly 90 years.

USCG Secretary-Class Cutters

The Coast Guard provides an interesting historical example of repurposing. In 1936, the Treasury Department built seven Secretary-class 327-ft cutters (Figure 6-3) modeled after the Navy's Erie-class gunboats. Their original purpose, envisaged during Prohibition, evolved into revenue cutters used for the interdiction of narcotics. Shortly thereafter, at the outbreak of the Second World War, they were rearmed and operated very effectively for the Navy in convoy escort duty and amphibious force flag ships. After the war, the USCG became independent again. The cutters were repurposed as weather ships and midocean search and rescue for transoceanic passenger aircraft. After rearming again, they performed coastal gunboat duty in the Vietnam conflict and returned afterward to midocean weather ship duties, until 1986. Because of their initial robustness, sea kindliness, and endurance, and the intentional repurposing, they served for half a century as "the Nation's maritime workhorses."[2]

[2]CAPT John M. Waters, Jr., USN (Retired). 1967. *Bloody Winter: Critical Months in the Battle of the Atlantic As Seen from the Conning Tower and Bridge,* J.D. Van Nostrand Company, Inc., Princeton, N.J.

FIGURE 6-3 USCG Secretary-class cutter. SOURCE: Courtesy of the Historic Naval Ships Association and the U.S. Coast Guard.

USS Enterprise (CVN-65)

The USS *Enterprise*, ordered from Newport News Shipbuilding in 1957, was the world's first nuclear aircraft carrier (Figure 6-4). She was in continuous service for over 51 years. From her original role as an anti-Soviet fighter plane platform, over the past half century she has deployed to provide strike support in the Vietnam and Southeast Asian conflicts, humanitarian aid, blockades, show-of-force in critical areas throughout the world, air support in Iraq and Afghanistan, and numerous other missions. To fulfill these roles, the platform has been adapted, reequipped, lengthened, and otherwise modified to meet the needs of new missions with new technology, aircraft, weapons, etc. This second oldest U.S. Navy commissioned vessel (decommissioned in December 2012) was repurposed numerous times because her size, robustness, and endurance capabilities made it possible. She has been able to quickly and effectively respond to surprises.

Spruance-Class Destroyers

The Spruance-class destroyer is another example of repurposing (Figure 6-5). This class was built during the 1970s to replace the Second World War

FIGURE 6-4 USS *Enterprise*. SOURCE: U.S. Navy.

FIGURE 6-5 Spruance-class destroyer. SOURCE: U.S. Navy.

Gearing and Sumner classes. Its original mission was to provide antisubmarine warfare (ASW) capabilities and escort duties for carrier groups. The Spruance class was built in a semimodular manner, and that made repurposing easier. Accordingly, during the 1990s, a large number of these vessels were updated by the addition of 61-cell vertical-launch missile systems and Tomahawk missiles. The USS *Cushing*, the last of the class, was also fitted with a 21-cell RIM-116 Rolling Airframe Missile (RAM) launcher on the fantail. While the last Spruance-class destroyer was decommissioned in 2005, it is an example of how a surface ASW/ escort vessel could be repurposed to a multipurpose, guided-missile destroyer.

Ohio-Class Submarines

Eighteen nuclear-powered, ballistic-missile submarines (SSBNs) of the Ohio class were built starting in 1976 (Figure 6-6). In the 1990s, based on new strategic arms limitation agreements, rather than retire the early Ohio-class SSBNs it was decided to reconfigure some of them as nuclear-powered guided-missile submarines (SSGNs). Starting in 2002, four of the class underwent modifications to their Trident missile launch tubes. They were modified to accommodate large

FIGURE 6-6 Ohio-class submarine. SOURCE: U.S. Navy.

vertical-launch systems (VLSs), whereby the tubes could accommodate clusters of Tomahawk cruise missiles, submarine-launched, intermediate-range ballistic missiles (SLIRBMs), submarine-launched, global strike missiles, operating equipment for special operations forces (SOF), countermine warfare packages, surveillance and reconnaissance sensors, and a variety of other payloads. The submarines can potentially accommodate future conventional cruise, ballistic, and boost-glide missiles. These reconfigured submarines are also able to deploy and supply SOF.

This highly innovative and cost-effective reconfiguring of the Ohio class has provided the Navy with greatly expanded capabilities at a relatively modest cost. This was possible because the Ohio class was originally built with adequate robustness, the ability to forward deploy for long periods, and with adequate size and space. It is yet another excellent example of the Navy's successful repurposing efforts.

Repurposing Payloads

The key elements of repurposing an asset to enable new capabilities for new missions include upgrades of data processing capability, guidance and navigation, and energy management. Processing capabilities have been following Moore's law since the 1970s and have enabled an explosion of products in both the military and commercial sectors from small, smart, precision weapons and unmanned systems to personal mobile communication devices such as smart phones and tablets. Advances in guidance and navigation have allowed miniature weapons with meter-level targeting accuracy and personal location systems tied to advanced schemes using mobile communication devices. Finally, new energy sources and management techniques enable systems to perform longer in stressing environments. Unmanned vehicles, drones, and remotely operated vehicles (ROVs) have all greatly benefited from these advances.

Naval forces too have greatly benefited from taking assets with existing capabilities and quickly upgrading them with new capabilities to respond to "surprises" in the operational environments. This was demonstrated with the preprogramming of the SM-3 missile to neutralize a failing satellite in a destabilizing orbit by shooting it down in an operation known as Burnt Frost. Repurposing was responsible for the procurement of air-to-surface missile capabilities using unmanned air vehicles during the Iraq War. In each case, existing platform and payload capabilities were minimally altered to provide significant new capabilities within a short period of time. This saved significant development and testing schedule time by leveraging established system performance capabilities. Furthermore, the appropriate level of regression testing was identified, which appropriately set testing and qualification requirements and focused them on the new capabilities to enable earliest deployment times.

Besides avoiding the usual requirement creep of a new systems development, repurposing eliminated the proposal development and evaluation cycle and potential protest delays and allowed requalification by similarity for certain subsystems—all saving time and money for DOD. While software modifications do not come free, upgrades can have significantly less impact on the test and qualification process than new hardware or systems replacements. Finally, with software modifications, impacts to the interface control documents for the platform interfaces can be minimized to facilitate deployment.

The committee advocates taking this same concept of software modification down to the hard subsystem and component levels. Building in excess capacity in a system will position it for future growth or added capability. Subsystems and components, such as firmware or even power amplifiers, for example, need to be designed with the ability to change functional performance without physical replacement. Using these key critical components as an investment or hedge for future capabilities, they can be leveraged as the building blocks to enable a quick turn to respond to future capability surprises.

This represents a change in the existing engineering design philosophies. The challenge will be in determining the proper balance between existing and known requirements and potential requirements down the road. The question to be answered is this: At what point does hedging our future needs with a single system start to drive the overall system development cost to the point where two separate systems may be more cost-effective? This will be the challenge for design teams of the future and will determine how they approach the allocation of resources to achieve system expandability, affordability, and agility in the face of capability surprise.

Limitations of Repurposing

While the benefits of repurposing can at times be huge, it must also be recognized that not every platform or payload lends itself to repurposing. Inadequate design margins, light scantlings, limited stability, lack of space or capacity, insufficient speed or endurance, and the like, may preclude adapting a platform or payload.

"Jumbo-sizing" or major conversions are sometimes a solution, but in many cases, responding to capability surprises by repurposing is not the right solution for reasons of cost and, especially, timeliness.

ARCHITECTURES

Concept

Several organizations interviewed by the committee described a regulation-burdened acquisition program (Figure 6-7) and said it was an almost insurmountable barrier to preparation and rapid technology response to any capability

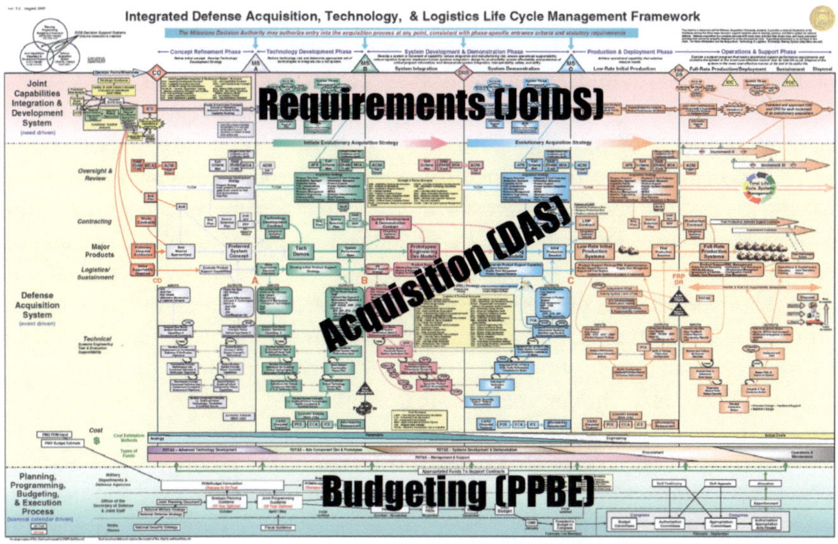

FIGURE 6-7 Framework process built for the risk averse. SOURCE: Jacques S. Gansler, University of Maryland, "Fulfilling Urgent Operational Needs," presentation to the committee, Irvine, Calif., June 27, 2012.

surprise. The committee recognized an even more foundational issue: that naval surprise normally occurs at the operational and mission level, while naval acquisition organizations and processes are centered on platform delivery. Several promising suggestions were raised during the committee's investigations. Consciously building capacity and capability reserves (software, hardware, and weapons) into platform payloads could be an effective way to achieve the agility needed to respond to surprise. This approach minimizes the changes to the capital-intensive investments in platforms, while focusing on the packages that actually deliver the mission capabilities, and it emphasizes incremental improvements that can be rapidly implemented. Another suggestion presented to the committee explored formalizing and resourcing a mission syndicate composed of (1) platform, sensors, and weapons research; (2) requirements; and (3) resource and acquisition organizations that together provide contributions in delivery of a particular mission's capability. This is an enhanced version of the OPNAV N95 coordination of a mine warfare enterprise and the naval laboratory warfare center concepts, where the syndicate lead is the holder of resources and "buys" mission platforms, sensors, and weapons from the providers. A mission-focus approach to acquisition may inspire an engineering approach that is more system-of-systems oriented and that could access a broad array of mission resources to anticipate and respond to surprise.

There is a foundational barrier in developing quick technical responses to capability surprise. Whereas most surprises affect capability to execute missions—for example, improvised explosive devices (IEDs) and air-to-air (A2A)—the Navy acquisition program is fundamentally whole-platform-centric—that is, the platform and most of its major payload systems are of new design. Separating the payload capability from the platform capability and the payload from their subsystems is an important approach to reducing the time line from development to operational deployment (F/A-18 F developed a major block upgrade of the airframe first and then the main radar sensor with the active array radar upgrade, with good results on cost and schedule). This should encourage innovation through the creation of adaptive solutions, drive down the cost of change orders during development, and shorten the time line for deploying new capabilities. In addition, it will also shorten the time that ships spend in port for maintenance and repair during overhauls.

The drive for solutions that meet 100 percent of a program's original requirements results in products that do not support the operator's need in the face of a rapidly changing adversary. The existing process consists of a stove-piped requirements process—stagnant procurement processes, inflexible budget, prohibitive reprogramming restrictions—all in a risk-averse culture that promotes adherence to the letter of a contract at the expense of providing the appropriate capabilities to our service members. A recent CNO article[3] supports this message. As discussed earlier in this chapter, separating the development and release of the platforms (or the "trucks") from the payloads (new capabilities) is key to both repurposability and the spiraling-in of new capabilities. Naval forces' platforms can be operational for many decades. However, to meet or create surprises, modular payloads are needed that can introduce new capabilities that take advantage of new systems and technologies in a timely manner, which includes test and evaluation as well as development. Repurposing and spiraling permit maximum leverage of previous testing and evaluation, minimizing regression test requirements and time to deployment.

Example: Littoral Combat Ship

There is an increased appreciation for the concept of payloads and platforms in today's Navy. The recent deployment of the newest littoral combat ship (LCS 2), the USS *Independence,* is an excellent example of the potential for efficiently, quickly, and economically reconfiguring vessels for urgent and emergent surprises. The LCS 1 and LCS 2 are shown in Figures 6-8 and 6-9, respectively.

While the LCS is not a heavy duty "truck," its littoral mission (inshore and shallow draft) and its high speed and agility dictate a lightweight hull. However,

[3]ADM Jonathan W. Greenert, USN, Chief of Naval Operations. 2012. "Payloads Over Platforms: Charting a New Course," *U.S. Naval Institute Proceedings* 138(7):16-23.

FIGURE 6-8 LCS 1: USS *Freedom*. SOURCE: U.S. Navy.

FIGURE 6-9 LCS 2: USS *Independence*. SOURCE: U.S. Navy.

its preplanned modularity makes it a versatile and very surprise-responsive combatant. Its mission packages can support ASW, gun mission module (GMM), surface warfare (SUW), mine countermeasures (MCM), maritime security module (MSM), SOF, helicopter operations, vertical launch packages, humanitarian operations, and many other missions. The LCS also demonstrates the discipline required for dimensional standardization, interconnectability and provision of adequate electrical power, coolants, air, computer control interfaces, berthing spaces for mission crews, and other support facilities.

These are starkly contrasted with several examples of elaborately outfitted single-purpose vessels that will be very difficult to reconfigure at a future date for different missions and surprises (e.g., SSBNs, Aegis cruisers and destroyers, and Avenger-class mine countermeasures ships). By separating the platform from the payload, rapidly changing weapon systems and electronics can be readily and more quickly adapted to surprises.

It is not economically feasible, however, to adapt a platform by replacing either it or its embedded systems each time a new mission arises. What is needed instead is to replace only the modular weapons, sensors, and unmanned vehicles payloads. The payload-platform approach has two requirements.[4] First, the design of the platforms must anticipate their future role or the fact that it may have to be adapted to some unknown role. The platforms must be generously sized and configured to provide sufficient space and ready accessibility to mission spaces. Second, the platforms should have built-in, sufficient electrical capacity, cooling water, ventilation, and other auxiliary services to support future missions. Main and auxiliary machinery must also be as modular as possible to permit adaptation to technical improvements or future capacity requirements.

The companion to this requirement is to make the mission packages for future weapons and electronics as modular as possible. The interfaces must be standardized to quickly and inexpensively offer as much "plug and play" as possible.

An excellent commercial shipping analogy has been the revolutionary emergence of containerization. Prior to the 1970s, merchant ships were customized to suit the principal cargoes and the trade routes for which they were employed. Today, throughout the world, the platforms, modern container ships, are austere and nearly identical, varying only in size. The payloads, the containers themselves, are standard across the globe. The containers move seamlessly between all modes of transport, from marine to rail to highway. The ubiquitous ISO 20- or 40-ft demountable, van-type container dry boxes are often intermixed with refrigerated containers, tank containers, "cattletainers," car carriers, foldable and fixed flats, and many other types of units.

The military has adopted modular containers for many missions, including communication units, diesel generator power boxes, communication containers,

---

[4]ADM Jonathan W. Greenert, USN, Chief of Naval Operations. 2012. "Payloads Over Platforms: Charting a New Course," *U.S. Naval Institute Proceedings* 138(7):16-23.

hospital operating units, machine shops, office and housing containers, and more. There may also be lessons to be learned from expeditionary force constructs.[5]

Modularity and Flexibility of Capital Ships

Although the committee has described the advantages of modular design and reserve capacity in ship classes such as LCS and Spruance, it notes that they are not the capital ships (most important warships such as aircraft carriers and battleships) that are able to fight in large-scale conflicts with near peers or in antiaccess/area denial (A2/AD) scenarios. The committee also observes that certain other flexible features are presently evident in capital ships.

First considered are the present Aegis cruisers (CGs). The Ticonderoga-class guided-missile cruiser is based on a stretch version of the Spruance class, taking advantage of the reserve space and power of the Spruance class. Originally built during the cold war to defend our battle force from large-scale attacks, the sixth ship of that class featured the then-new VLS, a modular design capable of launching a variety of missiles. The class has also featured standard missile (SM) evolutions, whose key features, such as the control link and aerodynamics configuration, were basically unchanged over decades, enabling many upgrades, now including the latest SM-3 Block IB ballistic missile defense (BMD) round, future rounds, and the newest air defense round, SM-6. Primarily from combat system software changes and minor interface changes, the Aegis cruisers have proved extraordinarily flexible in hosting combat system upgrades to meet the threats since the first-in-class initial operating capability (IOC) in 1983. The standard Tomahawk configuration has also enabled relatively straightforward introduction of block upgrades to that weapon. The downside of this class is that, beginning with core-memory-based Navy Standard UYK-7 computers, software capability upgrades through multiple generation computing and programming technologies, often with multiple baseline upgrades in development in parallel, have proven very expensive and time consuming. Efforts to provide open system architecture and standard computing platforms, even while upgrades continue, have proven difficult but are clearly important.[6]

The same can be said for the Burke-class guided-missile destroyers (DDGs). The production line for the latest flight DDGs continues, and the present capabilities are far beyond those of the original ship class, including the recent conversion to BMD capability. Again, the rapid manner in which an Aegis DDG was converted from BMD capability into a temporary antisatellite capability for

[5]In an earlier NRC report, it was recommended that "in long-term planning for future amphibious shipping, the Navy should consider the feasibility of a common ship design for assault, prepositioning, and sea-basing missions." See National Research Council, 1999, *Naval Expeditionary Logistics: Enabling Operational Maneuver from the Sea,* The National Academy Press, Washington, D.C., p. 31.

[6]CAPT Dan Meyer, USN, and CAPT John Geary, USN. 1998. "Aegis Computing Enters the 21st Century," *U.S. Naval Institute Proceedings* 124(1):39-41.

Operation Burnt Frost is impressive and indicative of the inherent flexibility for rapid change under certain circumstances.

Large-deck carriers, especially nuclear-powered aircraft carriers (CVNs) and amphibious assault ships (LHDs), have also proven very resilient and adaptable. Modularity is basically reflected in the ability to change over to new generation aircraft. The evolution of aircraft operating from these decks has allowed many generations of capability and a wide variety of warfighting, airborne surveillance, antisurface warfare (ASUW), and ASW capabilities.

Clearly, the excess capacity, modular approach can take a variety of forms, from the LCS and Spruance approach to aircraft evolution to software changes and standard weapons configurations. Still, further efforts to provide open software architecture and, perhaps, tailored acquisition processes as a result would strengthen the resilience and capability response time.

Requirements

The concept of open architecture and open systems development needs to be driven down in the design process to include subsystem and component elements. The commercial cell phone industry develops new products on major and minor upgrade cycles, where only one-third of the internal subsystems are new and it takes three minor upgrades before a "totally new" phone is developed, thus saving time and reducing cost and risk. The ability to repurpose a system is driven by the few key critical components that are its building blocks and will drive the degree of modularity (and therefore spiral capability) enabled by the overall system.

The ability of a system to offer the adaptability and flexibility required to neutralize an adversary's surprise is enabled by its architectural design. Open or adaptive architectures have been in vogue since the 1980s, driven by the rapid growth of software in systems development. The commercial industries, particularly the personal computer (PC) and telecommunications areas, have taken the lead in publishing interface control definitions that govern how anyone would utilize the industries' capabilities. The PC is a good example of an open and adaptive architecture, where IBM published the standardized PC reference design and interface standards and a whole industry of multiple suppliers, from Dell to Acer to Toshiba to HP, all built their desktop computers to be compatible with this architecture and interface standards. A more recent example is the Google Android smartphone operating system and architecture. The Android-based systems now hold 60 percent of the smartphone market and are used by six manufacturers.

There are examples of open systems in the military as well. Linux software was one of the first commercially open operating systems for computers. This concept was picked up by the military, which favored open interfaces for its weapons systems in such platforms as the LCS and its mission modules. Many system developers will claim their system is "open" as long as you work with them; however, a true open system has a published set of interface control documentation that any developer can use without going back to the original creator.

Complete interface control documents (ICDs) were essential to the LCS program being able to remain on schedule with the platform delivery when the non-line-of-sight (NLOS) payload was cancelled and a replacement weapon had to be found. This separation of platform from payload architecture allowed the LCS program to meet schedule and avoid cost impact associated with engineering change orders to the platform due to NLOS's cancellation.

The committee believes that these modular and open systems architecture concepts need to be driven down in the design process, meaning to the subsystem and component element level. The adaptability provided with open systems architectures needs to be considered in system architecture in all software and hardware development potentially including component-level designs. The component and subsystem designs form the crucial building blocks upon which future enhancements to the system can be added with minimal system impact and time out of service.

As existing software functionality/capability can be added without making tangible modifications to the systems today, this same approach needs to be taken with the hardware building blocks. This will require that the appropriate tradeoffs be made between (1) software and hardware requirements allocation and (2) a requirement not to be allocated only where it is most easily executed. A balance will need to be struck between the location of the functionality and the ability to most easily modify its functionality in the future. All of this must be done within the appropriate cost and schedule constraints of the existing baseline deliverable and future spiral upgrade capabilities. The DOD customer must participate in the allocation of resources for future contingencies as part of these architecture decisions.

The concept of the payload value chain is one that should be considered in the architecture designs for all types of capabilities development. The degree of integration for any platform with its payload is driven by the class of platform involved. The type of integration can range from tightly integrated systems on satellites to more loosely integrated systems on aircraft, ships, and submarines. The direction taken is often dictated by the mass fraction allocated to the payload as part of the overall system, which is directly related to the size of the platform—larger platforms permit increasing modularity from subsystem-level replacement (F-18 line-replaceable units [LRUs]) to completely stand-alone payload packages (LCS ISO containerization of payloads).

Once the level of modularity is determined, one can go even deeper in the system/subsystem design to compartmentalize the ability to introduce capability changes. Device reprogrammability enables other insertion points for design changes permitting updates to software and firmware loads at the subsystem levels while eliminating impacts on form or fit characteristics and minimizes system downtime.

Finally, there are instances where a payload, such as an antenna, is tightly integrated with the platform because of signature requirements, etc. This can involve significant downtimes to introduce new capability and should be pursued only after modularity and reprogrammability options have been considered as part of the overall capability's architecture design. The degree of smart modularity is architecture driven and should be a requirement for all new capability developments.

In all cases, it is important to understand the criticality of interface control documents (ICDs) to the ability to introduce new capabilities. Properly designed, extensible, and executable ICDs are the keys to enabling efficient, timely, and affordable introduction of new capabilities to existing fielded systems.

Software disciplines have evolved and enabled the efficient introduction of new capabilities with minimal impact to the hardware systems. Today's electronic designs have similarly matured to the point where the potential for in-place upgrades to hardware functionality are possible. As software design margins require spare processing and memory for future expansion capabilities, new electronic designs should similarly have (in-place) spare margin requirements to likewise support future capabilities.

Stealth Payloads, an Added Benefit

Clausewitz tells us that surprise is "the universal desire. . .basic to all [military] operations, for without it superiority at the decisive point is hardly conceivable."[7] He goes on to say that "the two factors that produce surprise are secrecy and speed."[8] This provides yet another reason for separating the payload from the platform. The prolonged pace of platform development and construction hardly provides the speed necessary for surprise. Ship design and construction programs as currently carried out thwart any attempt to keep them secret. However when payloads can be developed more quickly and can be designed, tested, and produced in a far more confidential environment, the likelihood of being able to produce or counter surprises is greater.

RAPID ACQUISITION

During the recent Iraq and Afghanistan wars, dozens of rapid acquisition organizations were created throughout the Services, including the following:

- Joint Improvised Explosive Device Defeat Organization (JIEDDO),
- Rapid equipping force,
- Quick reaction capabilities,

[7]Carl von Clausewitz. 1989. *On War*, edited and translated by Michael Howard and Peter Paret, Princeton University Press, Princeton, N.J., p. 198.

[8]Ibid.

- Mine-resistant ambush-protected (MRAP) (military vehicle) task force,
- Biometrics task force,
- Navy's rapid action teams, and
- Navy's rapid deployment and development process.

Almost 80 percent of the approximate $50 billion JOUNS funding between 2005 and 2009 went to two organizations—JIEDDO and the MRAP task forces.[9] Other procurements made by such organizations include unattended ground sensors, rocket-propelled grenade (RPG)-protection systems, and ISR assets. Although these organizations were charged with addressing JUONS requirements from the field, the "urgent" in JUONS is relative when one notes that it took the Navy a median total of 391 days when addressing its own JUONS requests.[10] This is hardly a time frame that would be considered sustainable for any "rapid" event. While these organizations filled the field requirements called out in JUONS, they often specified products rather than the specific requirement to meet the operational need of the field commander. They did, however, demonstrate the ability to rapidly field new capabilities to the warfighter using existing mature technologies or commercially available capabilities.

The downside of these organizations was that too often products were deployed to the field without the proper sustainment systems associated with training and support services. The tying of these rapid reaction services to supplemental funds further exacerbated the problems of long-term support and planning for these systems. While progress was made in the logistical areas since early in these conflicts as systems were integrated into existing logistical support systems, it did result in several early deployed capabilities not being utilized by the field operators.

In addition, since the early stages of these rapid acquisition organizations, the Services have improved their use of experimentation and operational exercises to better understand new system capabilities and how to develop the necessary tactics, techniques, and procedures (TTPs) to support their successful deployment, including maintainability requirements.

The dozens of rapid acquisition organizations that have sprouted up over the last decade to serve our needs in the Middle East are not a permanent solution to the acquisition challenge. These models, varied as they are, are not permanently sustainable in the current environment owing to existing laws, their reliance on supplemental funding, and the change of leadership in the executive and legislative branches. The combatant commander's (COCOM's) requirement for timely responsiveness to their needs reflects the need for a cultural shift in the acquisition process. For instance,

---

[9]Jacques S. Gansler, University of Maryland, "Fulfilling Urgent Operational Needs," presentation to the committee, Irvine, Calif., June 27, 2012.

[10]Ibid.

- The 80 percent solution may suffice in time-constrained situations.
- Testing should focus on what the system can and cannot do rather than on the pass/fail criterion of a requirement.
- Risk should be recognized and managed rather than avoided at all cost.
- Funding lines need to allow more flexibility for needs across the Future Years Defense Program.

Congress, DOD, and industry must change their current methods to support these objectives. Congress has to allow more flexibility into the funding profiles to address shortcomings in capabilities in months rather than years. DOD needs to move from a strict requirements-driven evaluation to a more managed-risk approach to providing timely solutions and must be willing to spiral in capabilities as they mature through the development cycle. Industry must be honest regarding its ability to deliver truly mature technology and systems on demanding time lines. Finally, the current unwillingness of the DOD to utilize foreign technologies and solutions, as well as proprietary solutions from industry, should be reconsidered. Given the pace of technology change, it is no longer appropriate for the DOD to assume ownership of technology and systems for which industry assumed the development risk.

The urgent needs of the COCOMs will continue to exist long after the current wars have concluded. In the face of the barriers identified above, there is little likelihood that a reversion to traditional acquisition methods and processes will meet the demands for "rapid" acquisition of solutions. What should be considered is a separate, rather than a parallel, acquisition agent to implement the changes necessary. Consistent with the philosophy expressed in the Clayton M. Christiansen's *Innovator's Dilemma*[11] and the recommendations in the Defense Science Board (DSB) report on fulfillment of urgent operational needs,[12] a separate acquisition organization should be established to address the barrier to rapid solution fielding. A separate and new environment is necessary to effect the changes necessary and to ensure that change does not become buried in the bureaucracy of traditional institutions.

Once a new organization has been established with flexible funding sources, the final step will be to staff it with the most innovative personnel. These people must be willing to think outside the box, manage (not eliminate) risk, and ensure that innovation is applied to the business and support systems to the same degree it is applied to the delivered product. As with the recent emphasis on foreign affairs officers, this career track must be seen as career enhancing and one that enjoys the commitment and support of senior leadership.

---

[11]Clayton M. Christensen. 2011. *The Innovator's Dilemma: The Revolutionary Book That Will Change the Way You Do Business*, Reprint Edition, HarperBusiness, New York, N.Y.

[12] Defense Science Board. 2009. *Report of the Defense Science Board Task Force on Fulfillment of Urgent Operational Needs,* Office of the Under Secretary of Defense for Acquisition, Technology and Logistics, Washington, D.C., July.

The process illustrated in Figure 6-10 and described in the DSB report supports the needs of both deliberate acquisition and urgent/emergent or "rapid" acquisition. The DSB's recommendation called for a new agency referred to as the rapid acquisition and fielding agency (RAFA). It should be a small, lean organization, with flexible funding and senior leadership sponsorship focused on speed to market for the warfighter.

Prototyping

Rapid prototyping, while a key component of accelerating technology and capability maturity, does not by itself go far enough to address the acquisition challenge associated with capability surprises. There is a need to take this a step further and deploy some limited numbers of new capabilities in order to allow system developers to engage in real-world experiments with the warfighters, to understand limitations, and create innovations in CONOPS to complement the new technology and to mature them consistent with the needs of the operational warfighters. Limited deployment also exercises the industrial production base and helps the operators to develop and improve the TTPs that are required to use the new capabilities.

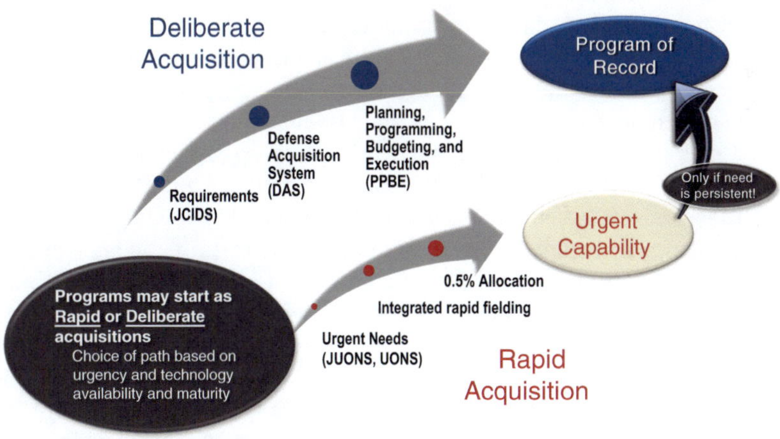

FIGURE 6-10 Dual acquisition path. SOURCE: Jacques S. Gansler, University of Maryland, "Fulfilling Urgent Operational Needs," presentation to the committee, Irvine, Calif., June 27, 2012.

Process

Historically in both the Second World War and in Iraq and Afghanistan, the DOD has provided relief from the Federal Acquisition Regulation System/ Defense Federal Acquisition Regulation Supplement (FARS/DFARS) regulations when the operational need was serious enough, resulting in both innovation and, at times, waste. At the same time industry responded with new developments and capabilities delivered to the warfighter in an expedient and efficient manner. Lessons learned from these experiences may provide the basis for a more timely but still cost-effective acquisition of capabilities in peacetime as well in wartime.

Current naval ship and weapon acquisition methods are not compatible with unanticipated surprises. Until the Second World War, given the robust industrial base and a sense of urgency as well as the benefit of the insulation provided by two oceans and the lower pace of attack, the United States was able to mobilize its industry in its own defense and also act as the arsenal for the world.

A shrinking world and modern technology no longer provide such luxuries. The current pace of naval vessel acquisition, up to 14 years from design to delivery, dictates that the old-fashioned approach of building ships and weapon systems to counter emerging threats no longer suffices.

Clearly, a viable acquisition system must be responsive to anticipated surprises. However, an acquisition system that is in step with unanticipated surprises must itself be versatile and agile. The acquisition of sensors, detectors, and weapons (payload) must be decoupled from the slow pace of ship and aircraft design and construction.

The deliberateness of naval ship acquisition means that "new" ships, from the time of concept design to commissioning, may span many surprises—technological, political, and economic. The usual result is that a ship, when it is finally commissioned, may not address new and emerging threats.

The current acquisition system is often criticized for its lethargy and complexity, fettered by FARS/DFARS procedures and oversight requirements. These in turn are further encumbered by budgetary and political considerations.

The acquisition system cries out to be streamlined to speed up and simplify the process, especially in its ability to be responsive to surprises and urgent needs. Other transaction authorities (OTAs) are moving in this direction. Many methods used in the commercial world for vessel acquisition, which are measured in months rather than decades for Navy acquisitions, are worthy of consideration. If the electronics development mirrors Moore's law, it suggests that quantum improvements in platform acquisition should be the goal. "We need to move from 'luxury-car' platforms with their built-in capabilities toward dependable 'trucks' that can handle a changing payload selection."[13]

The construction, outfitting, and manning of the LCS 2 also reflect an ap-

[13]ADM Jonathan W. Greenert, USN, Chief of Naval Operations. 2012. "Payloads over Platforms: Charting a New Course," *U.S. Naval Institute Proceedings* 138(7):16-23.

preciation for commercial practices. While clearly a naval vessel, many aspects of this vessel are common to merchant vessels.

The long-term pricing agreements (LTPAs) so prevalent in today's acquisition processes may be a challenge in an era without strong past performance evaluations due to low volume and competition from the commercial sides. This could very well lead to an environment where one is confronted with going to war with what is available, affordable, and on the shelf today, because the duration of the future conflict may not allow for hardware with new capabilities to mature, then be acquired and deployed.

Production

Simply developing and then placing a new capability on the shelf and waiting for a threat to materialize allows the industrial base and users to go dormant and creates a chasm with respect to future deployment—a challenge for the users and operators alike. It is akin to the setup times associated with a production run—things do not happen overnight, especially when surprise crops up. Our adversaries work this to their advantage operating inside our time line for response to take off-the-shelf capabilities and deploy them: While we are changing, they change their tactics yet again. The submarine force learned this lesson and kept the submarine design team engaged in the interim before the development of the Seawolf-class and the Virginia-class submarines so as to not lose the tacit knowledge of that community.

Our naval forces need to keep our adversaries at bay by constantly changing and showing new capabilities in relatively short periods of time. It may be more appropriate to demonstrate many small changes to capabilities (MRAP) than a few large ones (Aegis). While this will require a paradigm shift relative to how industry operates today, it will lead to a more agile and complex response able to rapidly produce many different capabilities in a much shorter time.

Our naval forces are no longer the sole possessor of technology, but to ensure that our systems behave as intended and without surprises, we must be able to trust their manufacturers in a world where the maker of a chip can tamper with its functionality and reliability, putting it beyond the reach of our system integrators and military operators. The United States needs to be exploiting the science from everywhere, leveraging technology development from our allies, and fielding systems from our own industrial base.

Our naval forces are a leader in technology development but no longer hold the dominant position they once enjoyed. With the advent of the Internet and the move to global supply chains, technology—and therefore capability development—is within the grasp of even the remote societies of the globe. Adversaries study our open military literature and are quick to devise simple yet effective countermeasures to our systems. Their proactive learning and understanding allow them to do this within the time frame of our present acquisition cycle. The observe, orient, decide, act (OODA) loop cycle of our adversaries has been shortened significantly because they no longer have to wait for systems to be used against them to learn how to counter their capabilities. Their reaction time is significantly reduced to well within our capability to react through our acquisition systems. "Rapid cycle of measure/countermeasure/counter-countermeasure will continue to add complexity to hybrid warfare operations, including cyber warfare."[14]

Spiral Development

> The Department of Defense's conventional modernization programs seek a 99 percent solution over a period of years. Stability and counterinsurgency missions require 75 percent solutions over a period of months. . . . Given the types of situations the United States is likely to face . . . it is time to think hard about how to institutionalize the procurement of [critical] capabilities and get them fielded quickly.[15]

Spiraling in capabilities is closely aligned with the three key elements of repurposability, described above. The challenge with spiraling in new capabilities is providing the expansion capabilities for the key elements in the original design. It is difficult to predict the future in any environment, and predicting "surprises" is no different. Spiral development requires discipline on the part of both the procuring agent and the contractor in laying down the relevant foundations for these elements based on reasonable expectations at the time.

It would be impractical to assume that one will be able to determine the exact amount of processing, memory, or power a future spiral capability will require. However, one can make reasonable estimates of technology progression and capabilities based on current technology and system trends. Providing a reasonable expansion capability based on these trends at both the component and the line-replaceable-unit (LRU) levels is the most important aspect of preparing for future spirals. The focus should be on minimal disruption to the physical aspects of the systems unit that provides the main functionality for the new capability.

This approach also calls for releasing incremental capabilities to the field as they become available throughout the development cycle in reasonable time frames. Adopting the model used by aircraft manufacturers to release operational flight programs (OFPs) to the wings on an 18- to 24-month cycle is a good example of spiral capability introduction. Early OFP releases contain fewer capabilities than later releases. However, they contain enough functionality for the

---

[14]Defense Science Board. 2009. *Report of the Defense Science Board Task Force on Fulfillment of Urgent Operational Needs,* Office of the Under Secretary of Defense for Acquisition, Technology and Logistics, Washington, D.C., July, p. 3.

[15]Robert M. Gates. 2009. "A Balanced Strategy: Reprogramming the Pentagon for a New Age," *Foreign Affairs* 88(1):28-40.

user community to gain valuable operational experience with the proposed new capabilities such that they can improve operational performance against an adversary's tactics and also provide valuable feedback into both the next spiral release and the accompanying TTPs. This enables fighters to more quickly introduce new and valuable (and lifesaving) capabilities to the field.

This approach calls for acquisition officials to work with industry to determine the appropriate release points for a spiral capability. In recent years, there has been a tendency to permit acquisition officials to drive the acceptance of a deliverable based on the letter of the contract. If spiraling is to be successful, both acquisition officials and industry will have to identify the point of "good enough," where sufficient new capability is available and useful to the operators such that it will make a difference in their ability to fight. This "good enough," capability becomes the basis for the operators to provide feedback to improve system performance and TTPs and to enable us to stay ahead of our adversaries by altering our tactics in a way that allows us to remain inside the adversary's OODA loop rather than the other way around. This ability to lean forward and retain the initiative rather than react to the enemy's tactics is a benefit of actively spiraling "good enough" capabilities to our warfighters in a timely manner.

The naval forces should deploy not with "deficiencies" but with "known capabilities" and spiral capabilities. The concept of establishing a baseline design and spiraling in upgraded capabilities has been around for decades for large platform systems such as aircraft. Where the platform and weapons system were tightly integrated, spiral upgrades were the best way to employ an initial new capability, even if it was somewhat limited relative to overall objectives, and then gradually improve or add capabilities over an 18-month block cycle. The B-2 aircraft (stealth bomber) is a good example. The aircraft was initially fielded without all the contemplated capability, and the aircraft was upgraded from Block 10 to Blocks 20 and 30 and now the Block 40 configuration is in the operational fleet. Even though the B-2 was a highly integrated design, it was architected to be deployed in incremental block configurations and some flexibility was designed in at the beginning of the program.

This worked fine in an era when the United States found itself controlling a particular aspect of the battle space, such as air superiority. What is needed today is to instill this same thinking into the even lower levels of our system development. As platforms are separated from the payloads, the payload system from the subsystems it comprises, the software from the hardware and further at the subsystem, LRU, SRU, and, finally, the component levels, the question is how to drive an 18-month block cycle (today the blocks last many years) down to several weeks (9-18 months would be a realistic and worthy goal). It also requires one to start thinking about our tactical/payload systems in a more strategic manner.

One needs to consider that our weapon systems may have a 50-yr life cycle, given the thinking that has kept the B-52 bomber in the inventory for three generations of pilots. As the naval forces can no longer afford to replace ships and aircraft every 10 to 20 years, this same approach must be brought to our tactical equipment in order to prolong their longevity through continuous spiral upgrades.

This approach to thinking about our tactical systems starts with the baseline acceptance: a recognition that the 80 percent solution is often "good enough." This approach offers two things: first, it allows the timely development of TTPs that will influence future cycle upgrades, and, second, it will allow us to evaluate the new capability's overall potential effectiveness in a more timely manner. This agility is required to keep the adversary on the defensive rather than to have us react to their threats. We now align our OODA loop more closely with theirs. This flexibility in system development and deployment and agility in responsiveness keeps the adversary guessing about our TTPs and how to react.

This change also requires a change on the part of industry. Industry needs to stabilize requirements at the 80 percent level while delivering new capability. It needs to develop flexible, modular system designs down to the component level if possible and demonstrate the ability to deliver on a block cycle lasting months rather than years. Finally, the military and industry need to set the risk/reward points to allow the flexible designs for system "repurposeability."

Rapid response capability will be the avenue taken when surprise happens—and it will happen regardless of one's planning. Naval forces need to learn how to deploy the 80 percent solution, not with "deficiencies" but rather with known "capabilities," and then learn how to spiral in capabilities quickly.

Examples of Rapid Acquisition Programs

The committee has identified several novel initiatives that have attempted to address the challenges of expediency with respect to the acquisition process. These initiatives include the USMC Combat Hunter program and the Navy's Acoustic Rapid COTS Insertion (ARCI) project and the P-8A Poseidon aircraft.

It is the intention of the committee that such initiatives to rapidly field new capabilities to counter unanticipated surprises should be separate from the existing process and not just incremental to it. This is necessary if the initiatives are truly going to help us to get new capabilities into the field in a shorter time frame for our warfighters.

USMC Hunter Warrior Series

The Hunter Warrior series was an outcome of General Krulak's vision of the future fighting environment that was forecast during his time as Commanding General of the Marine Corps Combat Development Command (MCCDC). He referred to the overall concept as Sea Dragon and fully implemented this plan when he became Commandant of the Marine Corps with the standup of the Commandant's Warfighting Laboratory. It consisted of three broad experiments which addressed the future warfighter challenges in the urban environment and small dispersed team operations: Hunter Warrior, Urban Warrior, and Information Warrior. The key elements enabling the delivery of operational capabilities within 24 months were (1) central experiment development, direction, and funding; (2) a small cadre force co-located at the warfighting laboratory; and (3) the use of operational forces to demonstrate the utility of new capabilities.

Hunter Warrior developed methods to increase the effectiveness and survivability of small dispersed forces on the modern battlefield. Within 24 months, inclusive of several intermediary milestones, MCCDC was able to develop, test, and deploy solutions to address mobility and communication challenges of the forces.

Specifically they produced an innovative solution for the replacement of the M151 Jeep. A commercial Mercedes was modified and deployed as the replacement vehicle and as an interim solution to the long delayed Fast Attack Vehicle (FAV) program, for which users had waited more than 10 years without a product. In the communication area, the program offered an alternative solution to the much delayed JTRS program by integrating small commercial handheld radios within the small units.

The Urban Warrior program investigated how to operate in the new urban jungle environment. It addressed the tactics, visibility, and first-respondent capabilities for the small unit fighters in this environment. It quantified the operational impact associated with supporting wounded soldiers and the need for improved uniforms and protective gear to improve warfighter protection. Additionally, it improved the MILES (laser tag gear) infantry combat training system with its predetermined types of combat wounds by introducing chalk rounds that identified the specific location and types of wounds Marines incurred during their combat operational training. Further, the program identified and transitioned immediately available commercial solutions to personal protective gear by adopting best practices for knee, arm, and other body parts, thereby minimizing the impact of cuts, scrapes, and the like on mission execution.

The Navy's ARCI Program

The ARCI program for the Navy's submarine force is an excellent example of how to deploy new capabilities in a short time. The keys to success for this program included a common baseline for combat systems across all submarine platforms in the fleet and the disciplined deployment of hardware and software updates to manage the risks associated with spiral innovations. Furthermore, the program recognized that advanced hardware development needed to be accompanied by advanced algorithm development. The release of these hardware and software improvements to the fleet in a staggered fashion, taking advantage of commercial practices and disciplined government component systems, and, finally, sea testing on a regular cadence is commendable.

The common combat system permits the Navy to leverage the associated costs across a relatively limited number of platforms in the inventory. Industry and junior and senior naval representatives collaborate on the spiral requirements and on determining the level of risk that can be safely managed on a particular spiral release. Commercial off-the-shelf (COTS) hardware is continuously updated on a 2-year cycle lagging the newest generation release in order to enjoy the benefits of observations made by early adopters. One-third of the fleet receives the hardware spiral every 2 years so that the entire fleet is upgraded by the sixth year. Software is updated on a yearly basis and trails any new hardware spirals by 1 year to ease integration challenges.

Tightly connected with these spiral capability releases is a contracting process that provides a steady budget line and allows for flexible contracting methods in support of these activities. Operating within the existing federal acquisition regulations (FARs), contracting officers understand and execute their authorities in support of the acquisition, testing, and deployment of commercial hardware on a time-critical time line that (1) maintains the capability deployment lines and (2) leverages state-of-the-art commercial designs and software updates that dovetail neatly with the fleet's identified needs. The contracting officers are a critical part of the spiral development team, and their ability to deliver innovative capabilities, within the allowable parameters of the FAR, is a critical part of ensuring a proper defense against capability surprise.

ARCI was conducted in a budget-constrained environment much like we are seeing today. In order to ensure contractor cost and schedule performance, the Navy continuously incentivized contractors to perform by leveraging a steady stream of innovation from the Small Business Innovation Research (SBIR) process. This prevented one company from enjoying a monopolistic position on the program. Senior leadership provided program management with the fortitude and backing to replace underperformers, both in industry and government, with others who were willing to dispense with overhead and infrastructure for focusing on deliverable product.

The flexibility provided by commercial, open-architecture hardware permitted alternatives in terms of algorithm or software products. More than once the Navy successfully replaced its software provider and still maintained technical, cost, and schedule performance.

When the ARCI program commenced, the Navy acoustic program office had experienced regular program cost and schedule overruns. At the same time it faced a real and growing threat to our undersea acoustic superiority and was operating under a budget 75 percent smaller than our cold war budgets. It was clearly being asked to do more with less. Today, using an 18-month block cycle, the Navy enjoys a 17-year record of on-time, on-budget delivery to the fleet.

P-8A Poseidon Aircraft

The P-8A Poseidon development is a good example of where the Navy leveraged commercial advances and practices to improve military product development times and save cost. The P-8A procurement utilized a traditional top-down Navy requirements process; however, it significantly leveraged the commercial investments in the Boeing 737 platform from which it was derived. This not only saved design and development cost and schedule, but also helped to focus operational test requirements because of demonstrated commercial performance.

The P-8A mission packages were significantly different from its 737 commercial counterpart, thereby requiring new development and test requirements. Several structural changes were required for bomb bay and bomb rack modifications unique to a military aircraft. These obviously required some development costs, but the ability to utilize the existing 737 structure as a baseline helped to bound the alternative solution set for consideration. These modifications still drove the need for full-scale static and fatigue test assets; however, the available commercial data in other common and mature systems, such as the landing gear and other areas, helped to minimize platform development costs.

Areas that experienced a significant leverage of the commercial 737 design included the engines and flight avionics hardware (software was a new development effort). Several mission systems were leveraged from other military systems to accelerate development efforts. These included electronic support measures (ESM) from the F-18, an acoustics package, and a repackaged radar from the P-3. Savings were realized in terms of both procurement (common supply chains) and the certification process requirements. Leveraging these commercial practices enabled the Navy to save one-third to one-half of the cost of having to develop a new platform from scratch.

It is significant that P-8A production is conducted on a third line in parallel with the existing two 737 production lines. This eliminated the start-up costs (in terms of schedule and dollars) associated with a new production line and was critical to controlling costs. This was due to the need to conform to existing practices already in place with the commercial line. The new military aircraft's development was heavily influenced by the commercial production rate line, which offered "infrastructure" already in place such as change review boards and process controls that drove behavior and thinking on the P-8A such that it did not impact the commercial production lines. Unlike previous military derivatives, the P-8A unique modifications are made in sequence during fabrication and assembly. This was difficult at first but later was recognized to accelerate the control and disciplines on the P-8A development, which in turn helped to control cost and schedule performance. Finally, the colocation of the military and the commercial production lines enabled the military to enjoy the benefits of commercial performance improvements (in connection with the aggressive continuous improvement margin targets) as they became available, further improving system and program performance, at a lower cost than if it had created those improvements itself. The drive to keep the commercial and military production lines as common as possible drove contractor and customer alike to implement these cost-saving measures.

In the testing area, the 737 certification data did little to eliminate developmental test and evaluation test points owing to the significant structural and mission package development. However, they did help to inform the test program decision makers, who were able to take advantage of them to focus required test points for the program. In the Navy Structures Group, where data must be created for each new platform, the available commercial data were leveraged to help the group make informed decisions about what to submit to M&S and what to actually test. Most of this leverage is skewed toward the development/requirements part of a program. In the operational test and evaluation (OT&E) portion, mission or operational performance drives the test schedule and must be accomplished for any new class of platform regardless of its commercial heritage.

As a multimission platform it was important for growth margins for all systems to be included in the program requirements. Weight, size, power, cooling, and processing, among others, all had specific technical performance measures (TPMs) that were set early in the program. As the program progressed, these TPMs were constantly reviewed and system trades were made in order to maintain growth requirements. One such trade involved the weight margin of the platform, where a more efficient engine, leveraged from its commercial 737 counterpart, was incorporated in order to preserve overall system performance (range at full load).

The above engine example typified the benefits of leveraging the commercial designs. Another involved the leveraging of the technical manuals (TMs) for maintenance and repair as well as operations. While Navy-specific requirements were added to the commercial TMs, the maturity of these documents helped the program accelerate its operational readiness and later enabled the Navy to move from contractor logistics supplied (CLS)-based maintenance to an organic-based logistics function more rapidly than originally envisioned.

In summary, the P-8A is a good example of leveraging commercial designs and practices to meet military needs in a timely manner. Utilizing the commercial 737 baseline, the Navy was able to realize cost and schedule savings during development, test, and maintenance that will reap benefits through the life cycle of the platform. Furthermore, the practices employed here offer lessons to consider when faced with delivering new capabilities in the event of a capability surprise.

TEST AND INITIAL TRAINING

Testing

Current test and evaluation practices are not taking full advantage of advancements in modern design, M&S, and coupon-type testing.[16] The earlier the involvement of the OT&E community in the development of requirements, while maintaining the appropriate level of separation required to avoid conflicts of in-

[16]"Coupon-type" testing refers to the use of a small piece of material for testing. These results may then be extrapolated into results for a larger, more costly piece of material.

terest, could reduce the time and cost associated with delivering new systems to the field in the face of capability surprise. With $30 billion of a total $70 billion OSD RDT&E budget dedicated to OT&E activities, including a Navy component of $13 billion, there is the potential for substantial savings that could be leveraged elsewhere.

The hypothesis stands that for incremental or spiral improvements to systems, as well as with new capabilities to address threats presented by surprise, the increased use of commercial data and practices can accelerate the fielding of these new requirements along a shorter time line. There is no longer a requirement or need to test full system articles until they fail or break completely. Best commercial practices leverage M&S analysis, coupon-type testing, and modern tools that are available to reduce the overall cost of such testing. The Navy has an opportunity to lead the other Services in this area and demonstrate the utility of such testing while enjoying the associated savings in time (and dollars) to operational deployment.

Past examples of this type of fielding include the Marine Corps Sea Dragon program, the space community's Mars Curiosity rover, and the Navy's P-8 program. In each of these instances, the use of the commercial data resulted in or offered the opportunity to, in hindsight, realize substantial savings in terms of schedule and dollars.

The Marine Corp Sea Dragon program, under the auspices of the Hunter Series of exercises, deployed improvements to mobility and communications capabilities by creating an integrated process team (IPT) of MCCDC, Systems Command (SYSCOM), and OT&E representatives that expedited the test and evaluation of new capabilities to ensure warfighter confidence at deployment.

The Mars Curiosity rover is a shining example of a system development and deployment where operational testing was conducted on selected parts in parallel with development activities. These practices should be applied to the DOD's OT&E execution to realize savings without sacrificing confidence associated with traditional verification testing.

Finally, in the case of the P-8 it is observed that while a commercial aircraft was modified to perform a military mission, a full-fledged traditional OT&E was still required by program management. While some significant modifications were made to the original design, was full OT&E required? It would be a valuable exercise to compare, with the benefit of hindsight, the original commercial data with data from the OT&E results to identify points where previous commercial testing could have been more effectively leveraged, resulting in schedule and cost savings to the program.

Training for Initial Capability

Basic proficiency training, not only for OT&E but also for initial operational capability (IOC), occurs well before the specialized training focused on mission readiness and is confined to core qualifications for basic readiness. Some of this basic training is now delivered in the form of distance-learning, with remote testing to validate proficiency.

In the commercial and academic sectors, it is common to use adaptive software techniques to introduce variation into tests for engineering and other technical certifications. This technique ensures that people cannot game the testing system itself and is also used to introduce surprise elements into the test. The latter technique helps organizations validate that students are not simply drilling and repeating by rote but instead understand underlying principles and are prepared to apply what they have learned to unexpected challenges.

In moving beyond initial training, naval forces could apply these same low-cost adaptive techniques to existing military distance-learning courses, adding capability surprise to the curriculum and, more important, to the distance-learning qualification tests. Once this testing regime has matured, surprise-related results from these tests could be fed into a broader U.S. Navy system managed by the recommended surprise mitigation office. Training is discussed in more depth in the following chapter.

Whereas more modern platforms are being designed for open computing architectures, retrofit of such architecture to legacy ships has been less successful. Some committee members recall that the original Aegis open architecture planning began in the 1990s, yet the transition to open architecture did not occur until late in the aughts (last decade) and then at considerable cost. As new computing platforms such as CANES are planned for combatant systems the committee is concerned that the open architecture of the near past represented by CANES could again become a constraint rather than an open architecture that is readily upgraded, given the long time lags between COTS equipment refreshes.

Rapid fielding of systems for naval mission needs was prevalent during the cold war. A program originally known as Battle Group Anti-Aircraft Warfare Coordination (BGAAWC) and then as Force Anti-Air Coordination Technology (FACT) was responsible for field testing prototypes on ships to evaluate such capabilities as radar detection and track automation, tactical link interoperability, and air track identification. Further, Space and Naval Warfare Systems Center (SPAWAR) would regularly field test capabilities to support C2 and communications connectivity improvements. There were some, however, who held that these systems were difficult to support in operation unless a full tooth-to-tail acquisition program was implemented. This was rarely accomplished because it was very expensive and would have taken a long time to achieve. Rather, rotating pools of equipment were provided and supported, some by contractors and some by in-service agents from the naval centers. By the late 1990s, as fleet systems became more complex and prototypes tended to be not well supported, a substantial slowdown in prototyping occurred that has persisted to this day. However, in this era of reduced acquisition and interest in rapid fielding, the committee believes rapid fielding of prototypes should be reconsidered. A tailored approach to in-service support for such rapidly fielded capabilities would be necessary.

FINDING AND RECOMMENDATIONS

Finding 5b: The Department of the Navy is not extending the full measure of open architecture principles throughout system development and deployment life cycles nor is it making best use of permissible contracting exceptions or best acquisition practices in responding to potential capability surprise in a timely and efficient manner.

Recommendation 5b: The Chief of Naval Research (CNR) should invest in discovery and invention (6.1 and early 6.2) research areas that take advantage of the entire payload value chain (i.e., payloads versus platforms; modularity versus integration; and reprogrammability), and inclusion of appropriate software and hardware design margins into development requirements. The Assistant Secretary of the Navy for Research, Development, and Acquisition (ASN RDA) should ensure that acquisition and contracting personnel are trained in the development of threshold versus objective requirements, the unique requirements associated with the use of commercial products, and the appropriate use of the waiver process in tailoring responses to potential capability surprise.

Recommendation 5c: The surprise mitigation office (see Recommendation 1) should encourage broader cross-organizational pre-planning in anticipation of, and based on previous, black swan events that can cut across U.S. government department responsibilities, and it should also serve as the lead resource officer for the rapid fielding of new capabilities to counter unanticipated surprises.

7

Force Response (Preparation and Readiness)

An important aspect of this study is how U.S. naval forces are preparing for response to anticipated and unanticipated capability surprises, as well as how they are developing and executing offensive strategies to surprise adversaries. This chapter discusses these topics by reviewing general preparedness, including quantitative and subjective measures of preparedness, and discussing what the committee has learned from its interactions with naval forces with respect to their current and planned actions.

FORCE READINESS—AN OVERVIEW

With regard to actual warfare, surprise is a certainty. It has been said that even the best-prepared battle plan is modified or changed with first contact with the enemy.[1] Operational and tactical surprise can take many forms and can vary in scope. The bottom line is that in addition to preparing battle plans against known and postulated threats, consideration and perhaps even anticipation must be given to the possibility of surprise capabilities in the hands of an adversary.

As part of this study, the committee reviewed preparedness and/or readiness to anticipate and counter these surprise threats. The DOD defines readiness as "the ability of United States military forces to fight and meet the demands of the national military strategy. Readiness is the synthesis of two distinct but

[1]"In preparing for battle I have always found that plans are useless, but planning is indispensable." GEN Dwight D. Eisenhower, USA (Retired) and 34th President of the United States, in Fred R. Shapiro, ed., 2006, *The Yale Book of Quotations*, Yale University Press, New Haven, Conn., p. 232.

interrelated levels."[2] There is individual readiness, which relates to the training, equipping, and performance capabilities of an individual, and there is unit/system readiness—the ability to provide capabilities required by the combatant commanders (COCOMs) to execute their assigned missions. "This is derived from the [estimated] ability of each unit to deliver the [wartime] outputs for which it was designed."[3]

Naturally, these definitions raise the question of how to measure preparedness or readiness. Several elements of readiness are quantifiable through set criteria such as equipment inventory, material status, personnel staffing, individual and crew training, logistics stocks, or adherence to directives. Other elements of readiness or preparedness do not easily lend themselves to being measured because they involve nonquantifiable judgments about leadership, unit/crew morale, personal interactions, mission execution, and the like.

Given that surprise is by definition unpredictable in its details but inevitable, it is unrealistic to attempt to prepare for all contingencies and surprises. As surprise is anticipated but unknown as to timing, scope, direction, and so on, a clear understanding of the mission and purpose of the operation by all personnel is important when preparing for unanticipated events.

A primary method of preparing, responding to, or adjusting to counter surprise is through additional guidance from the commander that emphasizes his concept and scope of the mission. This guidance take various forms, including commander's guidance, Commander's Intent, and Mission Concept of Operations (CONOPS); special procedures and actions called tactics, techniques, and procedures (TTPs); and rules of engagement (ROEs). (See Appendix E for explanations of the military terms in this paragraph and the next three.)

To improve our forces' ability to respond to and deliver tactical and operational surprise, our forces need to operate from a common training and operational base and they must have a thorough understanding of the mission at hand, a kit bag of TTPs, and a firm grasp of the ROEs.

In most instances the nature of the surprise, level of conflict, and ROE will provide a framework to determine which TTPs and urgent actions are appropriate for specific venues. Other factors might include these: Do the forces have access to the common operational picture (COP)? Do the forces involved have a shared local operational picture (LOP)? Are the forces involved governed by the same ROEs?

In general, the TTPs for regular warfare are straightforward and commonly practiced. However, TTPs to counter or respond to surprise in other types of conflict will frequently be less clear, and personnel may be hesitant to take

---

[2]Department of Defense. 2010. *Department of Defense Dictionary of Military and Associated Terms*, Joint Publication 1-02, November 8 (as amended through December 15, 2012). Available at http://www.dtic.mil/doctrine/dod_dictionary/?zoom_query=readiness&zoom_sort=0&zoom_per_page=10&zoom_and=1. Accessed July 11, 2013.

[3]Ibid.

timely action, despite the number of times such conflict has been seen before. Examples of such types include irregular warfare, military operations other than war (MOOTW), noncooperative target recognition,[4] information operations, and cyber operations.

Given the need for specificity in procedures under varying conditions, some examples of TTPs for an immediate response to counter surprise could include one or more of the following: emission control (EMCON), brevity codes, deception, deception events, limited access, cipher codes, quick reactions, disinformation, misinformation, operations security (OPSEC), countersigns, mike designator cards, and multiple option on-the-fly actions. More information about these is provided in Appendix E. Above all, once established, TTPs need to be routinely practiced.

TTPS AND CONOPS DEVELOPMENT FOR PREPARATION AND RESPONSE

Having TTPs is as important as having a CONOPS in order to deal with surprise. U.S. naval forces use exercises and training as well as experimentation to instantiate TTP and CONOPS and to develop and evolve them. However, today, naval activities are more typically devoted to planned mission areas, not potential surprises. Also, the committee found that experimentation programs involving new capabilities and new concepts have been curtailed.

Within the naval forces, specific organizations are responsible for developing TTPs and CONOPS. For the Navy, the Navy Warfare Development Command (NWDC), under U.S. Fleet Forces Command, directly supports fleet exercises and experiments. NWDC plays a central role in the development of guidance, doctrine, and CONOPS and also coordinates an annual experimentation program for the Navy. It engages in studies and analyses, modeling and simulation, war games, and red-team support and participates in exercises and experiments. It also collects experiences and lessons from recent operations. NWDC uses information derived from all these activities to evolve concepts, doctrine, and TTPs, all essential for adapting to surprise, though such activities are not typically

[4]Driving research into noncooperative target recognition (NCTR) is the fratricide issue, defined by MAJ Bill McKean, USA, as follows: The problem is our weapons can kill at a greater range than we can identify a target as friend or foe. . . Yet if you wait until you're close enough to be sure you are firing at an enemy, you've lost your advantage. The procedural approach of more restrictive rules of engagement (ROEs), according to McKean: What they found was, if you tighten the rules of engagement to the point that you reduce fratricide, the enemy begins inflicting greater casualties on you. Waiting until you're sure in combat could mean becoming a casualty yourself. Jim Garamone. 1999. "Fixes Touted to Combat Friendly Fire Casualties," *American Forces Press Service,* February 2. Available at http://www.defense.gov/News/NewsArticle.aspx?ID=41973. Accessed February 12, 2013.

directed at surprise. Other naval organizations also contribute to CONOPS and TTPs development.[5]

For the Marine Corps, the Marine Corps Combat Development Command (MCCDC) has overarching responsibility for TTP and CONOPs development. The command provides fully integrated Marine Corps warfighting capabilities, including doctrine, organization, training and education, materiel, leadership, personnel, and facilities. Experimentation activities are managed through the Marine Corps Warfighting Laboratory (MCWL), which provides guidance for concept-based experimentation for the development and integration of operational concepts and TTPs to enhance warfighting capabilities.

The Coast Guard established the unified Force Readiness Command (FORCECOM) with Service-wide responsibility for doctrine, training, exercise, readiness, and lessons learned and with centralized authority over all elements of the Service's readiness life cycle. FORCECOM develops TTPs and CONOPS in conjunction with existing doctrine, continuously testing operational guidance through interaction with specific FORCECOM units[6] colocated in key Coast Guard commands with special areas of responsibility (AORs).

MEASURING FORCE READINESS TODAY

How is force readiness actually measured today? There are many aspects to consider. As noted earlier, some are quantifiable using criteria such as equipment inventory and other material status, personnel staffing, individual and crew training, and adherence to directives. However, other elements are subjective in nature, such as leadership, unit/crew morale, personal interactions, mission execution, history and potential, etc.

For the quantitative aspects of readiness, one can look at known and measurable criteria. Assumptions can be developed and verified, and "status" can be measured, compared to desired levels, and charted to note current and forecast status of matrices for display and tracking can be developed. The result of this process provides insights but not a complete assessment of the preparedness or

[5]For example, the Fleet Forces Command (FFC) N03 unit analyzes capability shortfalls in warfare improvement programs to produce an integrated priority capability list and works with the training community to provide feedback continuously. The FFC Strike Force Training Atlantic and Pacific organizations extract lessons learned from prior deployments and inject them into the next carrier strike group (CSG) training cycle. The Naval Strike and Air Warfare Center (NSAWC), which is naval aviation's center of excellence, provides advanced naval aviation training and tactics development. NSAWC's charter establishes it as the authority on tactics, tactics development and training for all facets of naval aircraft power projection.

[6]The Coast Guard's FORCECOM executes through units located throughout the Service. These include the FC-A organization, which assesses readiness and in-place CONOPS and TTPs, and the FC-E, which specializes in supporting large-scale exercises. The FC-T has the majority of USCG training units that work in concert with the FORCECOM TTP unit (FC-P), which integrates and standardizes TTPs based on feedback from field units and lessons learned from exercises.

readiness status of an individual, a piece of equipment, or a unit. All of the naval forces currently have well-developed processes for identifying, measuring, and charting the quantifiable aspects of readiness.

The subjective aspects of readiness are much more difficult to identify and evaluate because they involve subjective judgments, such as the quality of leadership. These subjective readiness elements manifest themselves at the unit level as the mission is executed.

The Navy, the Marine Corps (USMC), and the Coast Guard (USCG) have implemented the Defense Readiness Reporting System (DRRS) as a direct replacement for the Status of Resources and Training System (SORTS), which had been used by all Services for over 25 years. SORTS focused on rules-based measurement or quantifiable readiness resources and people combined, measuring how they were trained and how the resources were allocated. SORTS for the most part did not use the commander's subjective assessment of the unit and graded its overall readiness based on the lowest individual rating or weakest link.

For this report, the committee will focus on the potential of the Defense Readiness Reporting System (DRRS-N) for the Navy and its use as a predictive tool. The USMC would be expected to use this reporting system in the same way. The USCG would only use this during national defense operations.

DRRS-N feeds the DOD DRRS-S, or Strategic Reporting System. DRRS-N is based on a common DOD framework that is capabilities based, is focused on missions, and heavily weights the commander's subjective assessment of the unit. It is Web-based and allows near-real-time evaluation of unit readiness. Readiness data is capability based rather than sortie based, as it is for the aviation community.

Mission essential tasks (METs), and a set of conditions for each task that is to be executed, are how commanders measure capabilities of their units. As an example, the guided missile destroyer (DDG) has 14 METs that the commander has to measure. The commander's assessment comprises the heart of the report, using his/her best evaluation judgment on each of the measures of performance.

Resource pillars that are reported on include personnel, equipment, supply, training, and ordnance (PESTO). The frequency of the report update varies by pillar. Personnel inputs are updated once a week; equipment and training are reported as they evolve but no longer that 30 days, and supply and ordnance are reported daily. Any significant event must be reported within 24 hours.

As an example, when a fighter squadron submits its report, it goes directly into the database; the air wing commander is not required to retransmit or collate. The Commander, Air Group (CAG) will report on his assessment of his staff and the overall health of the air wing, and that also goes directly into the database. The same applies for the strike group commander.

The Navy Readiness Reporting Enterprise System (NRRS) is the business intelligence or data aggregation group that interfaces with DRRS-N and multiple other reporting systems, including Aviation and Aircraft Carrier Readiness programs, Maintenance Supply Readiness, and Reserve Readiness programs, to name a few.

This is the system the higher echelon commanders use to manage the enterprise with business rules to compare reported and predicted values. Reports can be tailored to the needs of the commanders and staff. It used to have a capability search tool, but that has lapsed from lack of use.

Based on its review, the committee observes that the force readiness system is not being used to address the unplanned and the surprising. Most existing readiness assessments cover the range of missions and threats for which the units are currently tasked. However, surprises, by nature, may fall outside the current tasking of the units, so at issue is the degree to which the current readiness reporting systems can capture such outside-the-box situations. Surprise situations might be a result of a surprise during an armed conflict, a sudden humanitarian assistance mission, or an unexpected turn during diplomatic negotiations. An additional complicating factor is whether the "surprise" is of a kinetic or nonkinetic nature.

Present readiness reporting captures the current and forecasted status of several quantifiable readiness elements that have, over time, proven to have merit when measuring readiness against traditional missions and capabilities. Because many quantifiable skills and capabilities are transferable, such data could permit war planners and commanders to explore readiness, or gaps in preparation, for nontraditional and/or surprise events. However, this is not being done currently.

PREPARATION AND RESPONSE THROUGH EXERCISES, TRAINING, AND EXPERIMENTATION

It is clear to the committee that some important actions are being taken by naval forces to prepare for and respond to surprise. Dominating these are programs of exercises and training that include planning, conceptualizing, red teaming, and support of red cells. However, in such exercises and training, *existing* concepts of operation and TTPs are more typically applied. These activities take advantage of many of the provisions for surprise that have been designed into modern military equipment and capabilities but are not specifically tailored to potentially important but unexpected capability surprises.

Additionally, forces participate in a limited range of experimentation activities, some in conjunction with exercises, to enhance preparedness for surprise. Experimentation venues typically involve concept development, war games, modeling and simulation, and live events, many involving technologically advanced capabilities. Such exposure to and experience with innovation in an operational setting lead to new TTPs, as well as new technology, and serve to evolve concepts of operations. However, as will be discussed later in this chapter, such experimentation has been substantially curtailed, primarily due to current operational tempo.

Exercises and Training

The naval forces, numbered fleets, and combatant commands engage in substantial programs of exercises and training annually. For instance, the U.S. Pacific Command (PACOM) participates in more than 1,500 exercises and similar activities.[7] Such events typically have multiple objectives and often involve foreign military and coalition forces and assets. Major events can involve thousands of personnel and many platforms and capabilities. The Rim of the Pacific (RIMPAC) exercise for 2012 involved 22 nations, more than 40 ships and submarines, more than 200 aircraft, and 25,000 personnel.[8]

The exercises are an important component of preparation and maintaining combat readiness and hence reflect current U.S. strategic objectives and military strategy. For example, there is an increase in Pacific exercises anticipated in response to an enhanced U.S. military presence in the region.[9] The RIMPAC 2012 exercise was the largest one ever conducted (the exercises have been conducted every 2 years since 1971), in concert with that objective.

The Navy participates in hundreds of exercises annually, many involving operations with U.S. and multinational forces. The Marine Corps is currently expanding its exercise program for readiness including extending interactions with coalition forces, such as those of Japan and Australia, the latter to build coordination among amphibious forces. Future plans have Marine Corps permanently rotating forces through an Australian base camp and developing coordinated amphibious force capabilities.

The U.S. Coast Guard has more than its Title 10 missions to execute—for example, the safety of life at sea (SOLAS) mission area. Despite the daily tempo, commanders lead a broad array of unit-level exercises and rely on a regimen of exercises and training. Commanders have a continuous exercise program at the unit level when not prosecuting an actual mission and quickly divert to mission operations when the need arises.

Since the USCG has limited resources for exercises above the unit level, it focuses on large-scale exercises that represent the greatest likelihood for surprising and overwhelming Coast Guard operational forces. Examples are hurricanes[10] and spills of national significance. The USCG also participates in the biennial national-level exercises that involve most operational elements of national power.

---

[7]Available at www.globalsecurity.org/military/ops/ex-pacom.htm. Accessed February 12, 2013.

[8]Rim of Pacific Public Affairs. 2012. "RIMPAC 2012 Conducts Sink Exercise," *Navy News Service*, July 16. Available at http://www.navy.mil/submit/display.asp?story_id=68381. Accessed February 12, 2013.

[9]Michelle Tan. 2012. "Shifting Westward," *ArmyTimes*, February 20, p. 22.

[10]While Hurricane Katrina involved a scale previously unanticipated, a weapon of mass effect without intent and without decapitation of local government, USCG forces had already prepared for every element of the disaster.

Military Experimentation

Ongoing preparedness relies heavily on exercises and training but also on experimentation, though to a much lesser extent. Military experimentation by the Department of the Navy was examined extensively in a 2004 NRC study.[11] Much of the report is still relevant.

Military experimentation, per se, involves a spectrum of events, including studies and analyses, workshops, seminars and conferences, war games, modeling and simulation, as well as live events in the field. Experimentation is used to develop and evaluate doctrine, equipment, and TTPs—and, in effect, all aspects of the doctrine, organization, training, materiel, leadership and education, personnel, and facilities (DOTMLPF). Experimentation campaigns[12] provide a framework for learning about new capabilities and, as such, prepare for capability surprise.

Many naval organizations participate in experimentation, such as the numbered fleets, type commands, warfare centers of excellence, Office of Naval Research (ONR), and the acquisition community. However, as already noted, NWDC (for the Navy), MCCDC (for the Marine Corps), and FORCECOM (for the Coast Guard) play central roles in coordinating and/or managing programs of experimentation for purposes of evolving doctrine, CONOPS, and TTPs.

NWDC, in collaboration with Future Forces Command (FFC) and the Pacific Fleet (PACFLT), coordinates the current Navy program of fleet experimentation, called FLEX. The FLEX program manages and facilitates experimentation requirements involving services for the fleet and its assets. The recent 2012 events include an experiment in provisions for unmanned surface vessels to deploy nonlethal weapons, RIMPAC 2012 in Honolulu, Hawaii, the Trident Warrior 2012 in San Diego, California, and Valiant Shield 2012 in Honolulu, Hawaii.[13]

The U.S. Marine Corps has traditionally made use of experimentation to develop concepts of operation, doctrine, and TTPs. Recent operational tempo in Iraq and Afghanistan has significantly reduced the effort for experimentation. However, now the Marine Corps plans to increase experimentation through its exercise programs and through its Marine Corps Warfighting Laboratory (MCWL), which publishes the annual Marine Corps Science and Technology Campaign Plan. This plan describes the Marine Corps science and technology goals and objectives and the projected limited objective experiments.

---

[11]National Research Council. 2004. *The Role of Experimentation in Building Future Naval Forces,* The National Academies Press, Washington, D.C.

[12]An experimentation campaign is "a planned and cohesive, multiyear program of experimentation built on a series of experiments and related activities to develop the knowledge needed to inform major decisions about future forces, explore the viability of potential or planned changes to forces or their capabilities, and/or confirm that planned developments and directions will enable forces to perform as expected." From National Research Council, 2004, *The Role of Experimentation in Building Future Naval Forces,* The National Academies Press, Washington, D.C., p. 3.

[13]U.S. Fleet Forces Command. 2012. *FLEX News,* Vol. 1, March 26. Available at http://www.public.navy.mil/usff/fltexp/News_Media/201203_FLEX_News.pdf. Accessed February 21, 2013.

The USCG, though largely a law enforcement organization, has certain national defense and global responsibilities. It must deal with certain high-risk maritime capability surprises, such as preventing the implanting of mines in a U.S. harbor or a coastal attack on U.S. personnel or critical U.S. infrastructure through the use of semisubmersibles, small vessels, underwater unmanned vessels, or high-powered, sophisticated offshore container vessels. For its own preparedness, the USCG participates in joint experimentation and demonstration programs such as the Coalition Warrior Interoperability Demonstrations (CWIDs) and Joint Expeditionary Force Exercises (JFEXs).

SHORTFALLS IN CURRENT PREPARATION AND RESPONSE

Exercise and Training Programs

Today, naval forces conduct exercises and training to accomplish preparation. However, the committee reiterates that surprise is not being factored into these activities to the extent it should be.

The naval forces routinely engage in large numbers of exercise and training programs to achieve varying objectives, most of which are related to upcoming operational commitments. However, under such a regimen, there are few opportunities to inject real capability surprise and/or degraded environments into exercises and training.

With respect to exercises, opportunities may be increasing as overseas contingency operations (OCO) decline, but the current situation permits only limited excursions for surprise. There is little free play, and exercises are typically scripted with little deviation allowed.

This limits the naval forces' preparedness to practice dealing with surprise effectively and in a timely manner—and limits the development of CONOPS that would respond to anticipated and unanticipated capability surprises.

More representative of the current status is that deployment schedules and a rigid training regimen are constraining, allowing little or no time to inject surprise into exercise and training scenarios. There is, for example, little if any time to explore advanced or degraded operations. Even the time for basic proficiency training is limited. There are no empty blocks on the exercise and training schedules. Full deployment schedules and pre-deployment certification also leave little time for experimentation. Consequently there is insufficient time for concept development and for TTPs associated with the use of new technologies or new capabilities, primarily because of high turnaround and high operational tempo and basic training requirements. More ambitious training and experimentation for surprises would require curtailing current operational deployments, which might occur as OCO commitments decline.

There are indications that some exercises are proceeding that allow more extreme and/or degraded conditions and more "worst likely" scenarios. These provide anecdotal evidence of an emerging emphasis on anticipated surprise scenarios, such as a denial of space access, similar to the committee's surprise scenario on space in Appendix A. Examples of such exercises include Bold Alligator 2012 and Terminal Fury 2012, summarized as follows:

- Bold Alligator 2012 was the largest naval amphibious exercise in the past 10 years and demonstrated a revitalization of amphibious operations.[14] All of the naval forces participated. The exercise focused on revitalizing amphibious core competencies by looking to adapt skills to changes in force structure, technology, and culture. The exercise took place January 30 through February 12, 2012, afloat and ashore in and around Virginia and North Carolina, and involved over 20,000 U.S. and coalition personnel and 22 ships. What is particularly significant to this report is that this large 2012 exercise included degraded cyberenvironments and brought more information operations into play—more representative of realistic situations. Yet the exercise resulted in all forces having met all objectives, while "lessons were learned and gaps in processes and procedures were identified."[15] Recommendations were made to all components of DOTMLPF.
- PACOM conducts an annual major training exercise, Terminal Fury, to test command-and-control capabilities and prepare PACOM forces for western Pacific major contingency operations. The 2012 Terminal Fury exercise was cited in multiple briefings to the committee as representative of testing forces' response in a degraded environment, including one where space access is denied. It allowed practicing procedures enabling command, control, and communication (C3) without full reliance on space-based assets.

Other examples of exercises that allow degraded conditions and/or worst-likely scenarios include a recent Neptune Scissors event simulating carrier strike group (CSG) operations in an antiaccess area denial environment and the biennial national-level exercises involving naval forces, such as for hurricane preparation.[16]

While such exercises have been cited by some as examples of forces' preparation for surprise scenarios, they are not the norm. The constraints of training regimen, predeployment evaluation, and deployment schedules limit the resources available for excursions into surprise.

There are other factors that limit such excursions. For example, it is more

---

[14]Chad V. Pritt. 2012. "Bold Alligator Exercise Takes Fight to the Shore," *Navy News Service*, February 7. Available at http://www.navy.mil/submit/display.asp?story_id=65202. Accessed February 21, 2013.

[15]Expeditionary Warfare Collaborative Team. 2012. *Bold Alligator 2012 Final Report*, Condensed Version, U.S. Fleet Forces Command, Norfolk, Va., May 24, p. 6. Available at http://news.usni.org/2012/06/25/bold-alligator-2012-final-report. Accessed February 21, 2013.

[16]A biennial event involving all elements of national power focusing on capabilities to handle catastrophic events such as earthquakes, hurricanes, floods and fires, terrorist events beyond the 9/11 scale, and nuclear and biochemical warfare events.

typical to exercise while assuming chat rooms are operating and networks are functioning, because the denial of these would be "too hard." Not only are these situations complex to depict and to induce and analyze, they require substantially more resources and time than are routinely available for the exercises. Rather than ending a war game prematurely because of a disruptive breakdown in command and control, complicating and possibly prolonging a war game carries a resource impact. Continuing fiscal pressures will exacerbate the limitations on resources and therefore likely to continue to inhibit excursions. A scenario with a severely limited cybercomponent, for instance, is so disruptive that it necessitates that top cover. Some elements of the naval services have not trained to operate for extended periods in a denied environment for many years, though that condition was more typical some 20 years ago, when there was not such reliance on technologies like communications satellites and GPS.

Preparing for realistic cyberdisruptions is an example of emulating potential surprise that is especially challenging. Adaptation and emulation are compounded by changing network architectures, by fast-moving technology, and by new threats that emerge daily, such as attacks on trust certificates or new malware. In the real world, the effects are multiplied and scale upward quickly. Given emerging focus on better preparation by the U.S. Cyber Command (USCYBERCOM), the intent is to adapt training, red teaming, and exercises to better reflect what may be the real operational situation. However, to date progress has been slow. Yet the consequences of any cyber failure and penetration inside our command-and-control cycle would be significant. The methods to effectively protect and/or defend our own cyber and communications systems, especially at the lower echelons, are not well understood and not well funded. Additionally there is no shared understanding of an information warfare CONOPS.

Naval forces need to proceed more aggressively by using more with unpalatable but potentially realistic scenarios in exercises and training to establish and validate adequate procedures. Such procedures can include requirements for voice recognition and other forms of validation, which must function in less-than-optimal environments. Extreme and degraded scenarios test the envelope of capabilities, flag limitations and vulnerabilities, and foster the formulation of mitigation strategies.

In general, to become more ready to deal with these types of surprise, naval forces must more frequently train for such environments, including those reflecting nonkinetic attacks. This is not the norm today, and our forces must be prepared to deal with the nonlinear effects of surprise elements of many varieties. Too many exercises stop short of "real breakage."

Finally, there is a need to emulate our adversaries in exercises. This was emphasized in extensive discussions of red-teaming in earlier chapters of this report. The need for "realistic scenarios" includes the need to adapt and innovate like our adversaries. The committee discussed the issues of training against surprises such as space access difficulties and cyberintrusions with PACFLT. They concurred that more realistic exercise opportunities in these domains would be helpful. In fact, such exercise opportunities are beginning to be provided, but more time and resources are needed.

Military Experimentation

As noted earlier, military experimentation was examined extensively in a 2004 NRC study.[17] What is dramatically different from the state of play circa the 2004 NRC report and the present situation is the reduction in resources being applied to experimentation by the U.S. Navy and the USMC. The USCG was not addressed in the 2004 NRC report.

Experimentation efforts have been reduced dramatically primarily due to the operational impact of Iraq and Afghanistan. For instance, NWDC, in its central coordinating role for Navy experimentation under FFC, has experienced more than a 50 percent reduction in annual funding. At publication time, the 2004 NRC report cited funds available to NWDC for experimentation as totaling $20 to $40 million from ONR and $20 to $25 million from NWDC. Today that amount is approximately $15 million, which includes $14 million used for the FLEX program of experimentation.[18] This is approximately a 70 percent reduction in real terms.

The USMC, significantly smaller than the Navy in terms of resources, did not have substantial funding for experimentation in 2004; however it did operate effectively using selective and disciplined experimentation campaigns that were highly successful in transitioning new concepts of operation. Prior to 2004, the USMC had successfully accomplished a series of these campaigns, called Hunter Warrior and Urban Warrior, as described in Chapter 6. The campaigns were critical to building forces for urban warfare in Iraq and dispersed operations in Afghanistan by moving new concepts, doctrine, and TTPs to the field. Today the USMC has moved from multiyear longer-term campaigns toward small, limited objective experiments, such as those included in the RIMPAC exercises.

The USCG has limited funding for experimentation, beyond that for minor experiments at its Coast Guard Research and Development Center (CGRDC), and those experiments are not directed at capability surprise.

Military experimentation has changed. The ability to experiment at sea is currently sharply limited by the demands of the high operational and deployment tempo. Units are so busy with predeployment training and certification, it is difficult to find personnel or time on the training schedule to support a long range and centrally planned experimentation program. Additionally, the past decade of ground combat has created stressed naval forces because of their high operational

---

[17] National Research Council. 2004. *The Role of Experimentation in Building Future Naval Forces,* The National Academies Press, Washington, D.C.

[18] Navy Warfare Development Command, discussion with the committee on preliminary perspectives on capability surprise, May 16, 2012, Washington, D.C.

tempo. Budget concerns also contribute to the diminished service experimentation programs.

While forces strive to perform their assigned tasks, it is noteworthy that most of these tasks are conventional operations. However, surprise will likely come from an unconventional direction. Service-directed/coordinated and -funded experimentation will be necessary to prepare for just such an unconventional surprise.

Summary

There is insufficient time allocated to prepare for surprise in the training and exercise schedules, given certification and deployment requirements. Additionally, experimentation programs involving both the Navy and the Marine forces have diminished, while the USCG's reduced RDT&E budget has eliminated all but the most urgent small-scale experiments.

Overall, this lack of preparation for surprise exists for many reasons. While there will be a reaction to any serious surprise, the committee's concerns are whether the preparation and the response will be timely and effective, because the consequences of not being prepared could be catastrophic.

Additionally, in the committee's overview and its review of how readiness is measured, it has noted that capability surprise is simply not being taken into consideration today, even though there is potential for the current system to work.

The way forward to enhanced response to surprise by naval forces has multiple paths. One has to do with achieving an understanding of how to prepare and react effectively to capabilities by training with environments and scenarios that are realistic and representative of what adversaries could produce and induce. This path includes exercise and training time that allow local units the opportunity for excursions, and includes all types of surprise—including those may be self-induced, such as for conducting operations not currently envisioned.[19]

To summarize, operational naval forces are not preparing adequately for surprise in current exercise and experimentation approaches. Exercises do not usually allow degraded environments or unexpected developments to be realistically addressed. Training now focuses on current missions and operations and leaves inadequate time for experimentation and the use of new technologies. There is insufficient development of concepts and tactics for addressing surprise and of practicing to deal with their impact. Additionally, readiness metrics focus on current missions and do not address surprise.

Operational commanders should incorporate, when feasible, degraded environments and aspects of surprise into exercise and training scenarios to improve preparation for and response to surprise. The FFC, including directed type

[19]An example would be that of the United States abruptly deciding to conduct unrestricted submarine warfare against Japan in the Second World War when the U.S. submarine forces had no experience, weapons, or TTPs for attacking merchant ships.

commanders, MCCDC, and FORCECOM, should expand experimentation and related activities to develop concepts and tactics to counter surprise. To offset the limitations of scarcer resources, experiments can be designed on a small scale.

These commanders should use the results from exercises and experimentation to analyze and assess preparedness for capability surprise. As appropriate, they should formulate and incorporate measures into the existing readiness reporting structures through the appropriate naval organizations and systems.

The results derived from incorporating surprise into exercises and experimentation will be forwarded to the appropriate Service organizations, including the surprise mitigation office, and entered into the training continuum, as appropriate.

Finding 6a: U.S. naval forces are not preparing adequately for potential capability surprise in current exercises and experiments. For example, naval exercises do not usually accommodate degraded environments or unexpected developments to be realistically addressed, and training tends to focus on current operations and leaves inadequate time for experimentation and use of new technologies.

Recommendation 6a: Operational commanders should incorporate, when feasible, degraded environments and aspects of surprise into exercise and training scenarios to improve preparation for and response to surprise. U.S. Fleet Forces Command (FFC), including directed type commanders, Marine Corps Combat Development Command (MCCDC), and U.S. Coast Guard Force Readiness Command (FORCECOM), should expand experimentation and related activities to create concepts and tactics to counter surprise. To offset resource impacts, activities of limited scope, such as small-unit or small-scale experiments, may be utilized.

These commanders should use the results from exercises and experimentation to analyze and assess their preparedness for capability surprise. As appropriate, they should formulate measures and incorporate them into the existing readiness reporting structures through the appropriate naval organizations and readiness reporting systems.

The results of incorporating surprise into exercises and experimentation will be forwarded to the appropriate Service organizations, including the capability surprise office, and entered into the training continuum, as appropriate.

STRATEGIES FOR IMPLEMENTAION

The committee believes that there are some strategies to implement Recommendation 6a that do not levy extraordinary resource requirements or require substantial changes to existing systems and methodology.

To Increase Surprise in Exercises and Training

Commanders could use theater-specific surprise scenarios when the operational tempo is slow. As an example, NWDC could provide a series of specific scenarios—one might be denial of Suez Canal transit rights. This could be a tabletop exercise whose results are fed back to NWDC for processing and data mining when completed. This strategy has the benefit of exercising intellectual capital based on actual readiness and preparing operational commanders for potentially relevant occurrences. Such surprise excursions could also easily include specific aspects when representative scenarios brought up in this report—space access denial, social media manipulation, and the Fukushima disaster. The latter two have particular relevance because they resemble recent surprise events, such as the attack on the U.S. consulate in Libya and the disaster response required for Hurricane Sandy.

Military distance-learning courses could incorporate capability surprise. Basic proficiency training occurs well before the specialized training focused on mission readiness and is confined to core qualifications for readiness. Some of this basic training is delivered in the form of distance learning, with remote testing used to validate proficiency. In the commercial and academic sectors, it is common to use adaptive software techniques to introduce variation into tests for engineering and other technical certifications. This technique ensures that people cannot game the testing system itself and is also used to introduce surprise elements into the test. This latter aspect helps organizations validate that students are not simply drilling and repeating by rote, but instead have understood underlying principles and are prepared to apply what they have learned to unexpected challenges. Naval forces can apply these same low-cost adaptive techniques to existing military distance-learning courses, adding capability surprise to the curriculum and, more importantly, to the distance-learning qualification tests. Once this testing regime has matured, surprise-related results from these tests could be fed into both the broader and appropriate naval systems and into the capability surprise office.

Surprise, such as caused by our own processes, systems, and materiel is addressed by operational forces today through immediate action emergency procedure exams and casualty exercises. These are the norm in aviation, ship, and submarine units. Capability surprise could be practiced in the same manner.

To Expand Experimentation

Recommendation 6a cites the use of limited venues for experimentation. As noted earlier in this chapter, the use of military experimentation campaigns has been effective in the past. These consist of a full spectrum of planned activities and various limited experimentation venues, including war games and limited objective experiments, all supported by careful studies and analyses and careful, planned incremental progress. Such methods deliver systematic results and do not necessarily require large-scale events involving substantial fleet assets. The committee supports recommendations of an earlier (2004) NRC study[20] with respect to expanded use of such campaigns to maximize the effectiveness of experimentation and minimize the impact on resources. This approach could be effectively used to address surprise.

One example would be to plan a campaign of experimentation to evolve new TTPs, doctrine, and capabilities to combat and respond to different forms of attack that cause denial of space access under varying conditions. Potential events such as these are described in the first scenario of Appendix A.

To Measure Preparedness for Surprise

The current system, DRRS-N[21], has utility for capability surprise given the currency of quantifiable and commanders' subjective assessments of the units. The NRRE system could provide commanders an assessment for a variety of worrisome potential surprises. If, say, a humanitarian aid/disaster relief (HA/DR) scenario occurs in the western Pacific, including nuclear contamination, commanders would want to know the status of forces in the area of responsibility (AOR) and how they might contribute to the effort. A profile already created would search the NRRE database for a suitable HA/DR response capability. It would produce, in stoplight form, the status of all forces in the AOR giving a capability report on helicopter availability, small boat readiness, medical supply levels, medical personnel available, dosimeter availability and type, reconnaissance assets, supply reports and food stores, ability to generate emergency power and make potable water, ships availability, and fuel reserves, to name a few. These preset profiles could include any number of potential scenarios.

The merit of this approach is that operational units require no additional effort or reporting. Important potential surprises, i.e., HA/DR, denial of space access, noncombatant evacuation operations, would need to be staffed and defined at the FFC level. Specific metrics would then be identified and labeled for individual surprise profiles within the NRRE system.

A more ambitious approach, but more burdensome on reporting units, would be for FFC to define a reporting goal for the time nonengaged units should spend preparing for undefined surprises that the units themselves can identify. This would then be reported for inclusion in a database.

Since capability surprise is not integrated into daily naval forces activities and planning, it is essential that doctrine and training, especially for the commander, be established to ensure that leaders know how to exploit the information in DRRS. DRRS and NRRE are robust enough to incorporate additional data,

---

[20]See, for instance, Recommendations 3 and 4 in National Research Council, 2004, *The Role of Experimentation in Building Future Naval Forces,* The National Academies Press, Washington, D.C.

[21]As noted earlier, the USMC would use this system analogously, and the USCG would use it for national defense operations.

if required. Such information will, for the first time, give commanders useful measures of unit readiness for surprise in the context of specific mission areas.

Finding 6b: U.S. naval forces do not have an advocate and resource sponsor to rapidly field new capabilities to counter pop-up surprises, nor are they taking advantage of any existing capabilities, as identified in the Navy Readiness Reporting Enterprise (NRRE), that could potentially counter surprises of all types.

Recommendation 6b: Commander, U.S. Fleet Forces Command (FFC), should leverage the Navy Readiness Reporting Enterprise (NRRE) and provide operational commanders with any existing capabilities that could counter surprises of all types.

MAXIMIZING THE IMPACT OF OUR SURPRISE CAPABILITIES

Long-term scientific research, technological developments, and/or capability acquisitions enable both defensive and offensive surprise. These are discussed in earlier chapters of this report.

In reviewing such long-term endeavors through the lens of naval force response, the subject of this chapter, the committee has already noted the need to increase the naval forces' development and use of innovative capabilities, such as through enhanced experimentation programs. However, it also believes that our naval forces can improve their ability to spring their own capability surprises on our adversaries.

A number of surprise mitigation capabilities would be highly classified. The transitioning of highly classified developments and acquisitions into operational use can be problematic owing to the restrictions of "need to know." This can present a fundamental barrier to ensuring readiness to apply such a capability in event of a surprise. The following is a scenario that the committee believes may occur:

> A capability is under development. It would surprise an adversary and ensure an impact, but details necessary to ensure appropriate use are highly classified, often involving compartmented security constraints. There may be a general awareness of it by military operators, but there is extremely limited knowledge of scope and characteristics. It may, for example, involve a technology not usually associated with or demonstrated in military applications. CONOPS could, as a result, be thin, or scarce, or even nonexistent. Because there has been limited interface with the operational community and little to no experimentation involving that community, doctrine, training, and TTPs are missing or sketchy.

Consequently, the use of the capability for the designated military operation may be set aside or even forgotten because of its unknown consequences, or because its application is not sufficiently understood.

This possibility is inferred based on previous known examples. The next section is a well-documented example to further illustrate this difficulty and understand the potential impact.

An Example of the Difficulties of Accomplishing Surprise

The classic example of difficulties in bringing the Navy's (then) very secret new torpedo influence exploder to bear on Japanese warships at the start of the Second World War remains instructive.[22]

During the 1930s, both Japan and Germany ignored or circumvented existing treaties that limited warship construction and built a growing number of large combatants that were well armored above and below the waterline to protect them against the then-prevalent contact-fused anti-ship torpedoes. The U.S. Navy, concerned about this development, embarked on a highly classified program to develop a torpedo exploder that would respond to the magnetic influence of a target ship with the intent of setting torpedoes so equipped to explode under the target's keel.

The effective use of such a revolutionary new torpedo would indeed have been a major surprise for any prospective adversary. To protect this potential for surprise, the Navy tightly restricted knowledge of its existence and went so far as to ship to the fleet its inventory of the new exploders in sealed black boxes with no information on their prospective use. After Pearl Harbor, attempts to install the new exploders went awry, and the lack of direction for the proper employment of the modified torpedoes compounded the problem, as did the fact that the exploders had not been fully tested, in part to preserve security. It was of little comfort that the British and French navies were having similar difficulties with their own secretly developed exploders.

The net result was the loss of opportunity to take advantage of a major technical advance that would have been very effective against Axis warships had it been properly tested and employed. Instead, many allied lives were needlessly lost attempting to employ an ineffective weapon improperly, and the allies were forced to switch to contact-fused torpedoes for a long time early in the war.

To summarize, some key classified capabilities may not be disclosed to planners or operators and therefore will not be routinely incorporated into combatant plans or practiced by operators.

[22]Clay Blair, Jr. 2001. *Silent Victory: The U.S. Submarine War Against Japan,* Naval Institute Press, Annapolis, Md.

Finding 6c: It is unclear whether some key classified capabilities—to the extent any exist—are disclosed to planners or operators and therefore may not be routinely incorporated into combatant plans or practiced by operators.

Recommendation 6c: Finally, operational commanders should work to ensure that any of our key classified capabilities—to the extent any exist—are disclosed to planners or operators so that they are incorporated into combatant plans or practiced by operators in responding to capability surprise.

FINAL THOUGHTS

The committee believes that naval forces' preparation and readiness for capability surprise is insufficient. The current operational tempo has been a primary detriment to preparedness. However, pending budget cuts could significantly degrade readiness further. The committee believes that some actions, as specified in the Recommendations, may not require extensive expenditures of resources but will nonetheless improve the readiness posture. How these could be achieved is explained anecdotally in this chapter, with some examples linked to this report's representative scenarios.

8

Putting It All Together

Each of Chapters 2-7 provides details of one of six primary phases of effective capability surprise mitigation. As noted throughout, there are many nuances in each phase, but the overarching goal is to provide senior Navy leadership with an accurate picture of the landscape of potential surprises that naval forces may face. The chapters also discuss the available mechanisms to prepare for delivering a surprise or responding to one whether intelligence-inferred or disruptive, and whether mitigation would consist of defensive counters or offensive preemptive surprises of our own.

A number of surprise areas are currently being addressed by the Navy establishment, including by the Office of the Chief of Naval Operations (OPNAV), the Office of Naval Research-Global (ONR-G), fleet combatant and component commands, among others. Yet while each of these groups is addressing some aspect of surprise preparation, providing input through their respective chains of command, there is not a focal point within the naval forces enterprise to look holistically across the breadth of potential surprise or at the phases of response and preparation. No particular component of the Navy is able to advise the Chief of Naval Operations (CNO) and other OPNAV components of the institution's ability to address surprise and, specifically, certain intelligence-inferred surprises, especially those in the gaps between missions and programs, for which there should be some plan available in advance.

Naval forces can leverage existing capabilities and work to integrate the potential contributions to adequately prepare for capability surprise and also to ensure collaboration with the other services and with the Office of the Secretary of Defense (OSD).

As discussed in detail throughout this report, the committee has identified

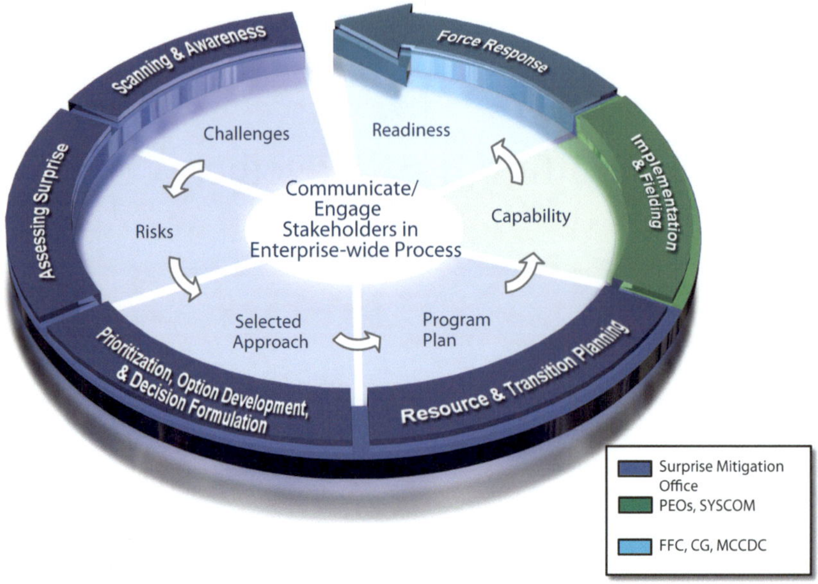

FIGURE 8-1 Recommended functional framework for addressing capability surprise.

six functional phases to be performed by a capability surprise mitigation office. These phases, illustrated in Figure 8-1, provide the backdrop for engaging the full array of capability acquisition, fielding, and training entities. To effectively coordinate and accomplish these tasks, a surprise mitigation office should be established within OPNAV. It could be a new office or it could also be an existing office with a modified charter within one of the codes. As discussed in Chapter 1, of the alternatives, the committee considers N9I the best option based on its present role, but recognizes that the CNO may select an alternative as better aligned with priorities. This office can serve as the focal point for capability surprise among all naval forces (i.e., the Navy, the Marine Corps, and the Coast Guard). While the office may not be the lead for all aspects of the capability surprise enterprise, it should serve as a leader element to ensure completion and delivery of any operational capability that addresses surprise mitigation.

The six key functions that should be represented in this office are as follows:

- Scanning and Awareness,
- Assessing Surprise,
- Prioritization, Option Development, and Decision Formulation,
- Resource and Transition Planning,
- Implementation and Fielding, and
- Force Response.

The preceding chapters describe these phases in some detail. In this chapter the committee tabulates their attributes. It then illustrates how they can be interwoven into the existing naval organization structure. Finally, the committee summarizes how the framework could be applied to more effectively address the three examples of surprise that recur throughout this report.

ROLES AND ACTIVITIES

Figure 8-2 summarizes the outputs, owners, stakeholders and participants, and activities of each phase of the framework as described in the previous chapters. The committee has determined that many, or most, of the requisite functions already reside within the naval organization. However, the functions are not sufficiently integrated, prioritized, or advocated. That is the primary motivation for Recommendation 1, which suggests that a surprise mitigation office be established to lead in coordination and prioritization. The committee shows the surprise mitigation office is the coordinating "owner" for the first four phases. The final two phases of the framework would be led by the existing organization once OPNAV has prioritized, defined, and planned the appropriate measures. OPNAV will continue to participate in the final phases and to ensure that the surprise mitigation capabilities are deployed in a timely manner. The stakeholders and participants are brought together, as appropriate to the topic, to address signs of emerging surprises from the scanning and awareness activities of the operational, research, and intelligence establishments in phase 1, and to move toward the modeling and assessment organizations and laboratories to verify feasibility in the middle phases. Finally, the acquisition and operational organizations are the primary players in the final phases. The activities of each phase are summarized, and the primary output of each phase is shown in the first row of the table, from identification of challenges to delivery of capability and readiness to deploy. The committee notes that sometimes, with disruptive surprises recognized only in the heat of an operation, a crash program will involve just the last three phases as the early ones have been preempted by new findings on the battlefield or disaster area.

ORGANIZATIONAL ALLOCATION OF STUDY RECOMMENDATIONS

Because this study is largely about the integration of leadership and processes, the recommendations are necessarily focused on different parts of the naval organization. The recommendations must be viewed as an integrated set—that is, implementing some recommendations, but not others, will not lead to an integrated process in the committee's view. On the other hand, the committee worked to ensure that the present organization is leveraged to the maximum extent and with a minimum of change.

For the convenience of the reader, Figure 8-3 is provided to map the recommendations to elements of the naval organization, and Box 8-1 summarizes the

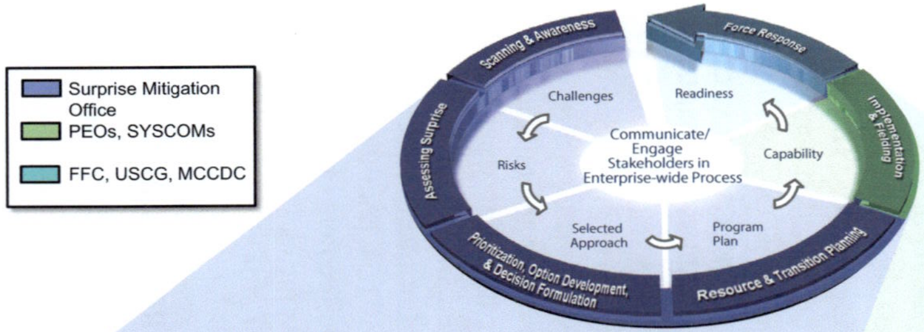

	Scanning & Awareness	Assessing Surprise	Prioritization, Option Development & Decision Formulation	Resource & Transition Planning	Implementation & Planning	Force Response
Outputs	Challenges	Risk	Priority, feasibility, & affordability	Program plan	Capability	Readiness
Owner(s)	Surprise Mitigation Office	Surprise Mitigation Office	Surprise Mitigation Office	Surprise Mitigation Office	PEOs SYSCOMs	FFC USCG FORCECOM MCCDC
Stakeholders	ONR-G ONI FFC USCG FC-A MCCDC	N81 FFC USCG FC-A MCCDC	N2/N6 N8 N9 DASN RDTE/CTO ONR/CNR MCCDC USCG FC-P	DASN RDTE/CTO MCCDC USCG CG-8/ FORCECOM SYSCOMs	Surprise Mitigation Office N9 N2/6 DASN RDTE/CTO MCCDC USCG FC-P SYSCOMs	OSD/SCO Surprise Mitigation Office USCG FC-E & FC-T
Participants/ Performers	Industry Labs/Warfare Centers Academia IC	NWDC, MCWL Industry Labs/Warfare Centers Think tanks Services' Colleges	NWDC, MCWL Industry Labs/Warfare Centers Think tanks Services' Colleges	Industry Labs/Warfare Centers PEOs SYSCOMs ONR	OPTEVFOR Industry Labs/Warfare Centers	USN, USMC, & USCG Operational Forces
Activities	Technical Intelligence S&T monitoring Physical modeling	Campaign models Cultural assessments System modeling	TTP Modeling Prototype Experiments/ATD	Program Planning, Budgeting, & Coordination	TTP/CONOPS System Modifications New Programs	TTP/CONOPS Assessment Testing Training Exercises Equipping

FIGURE 8-2 Roles and activities to address capability surprise.

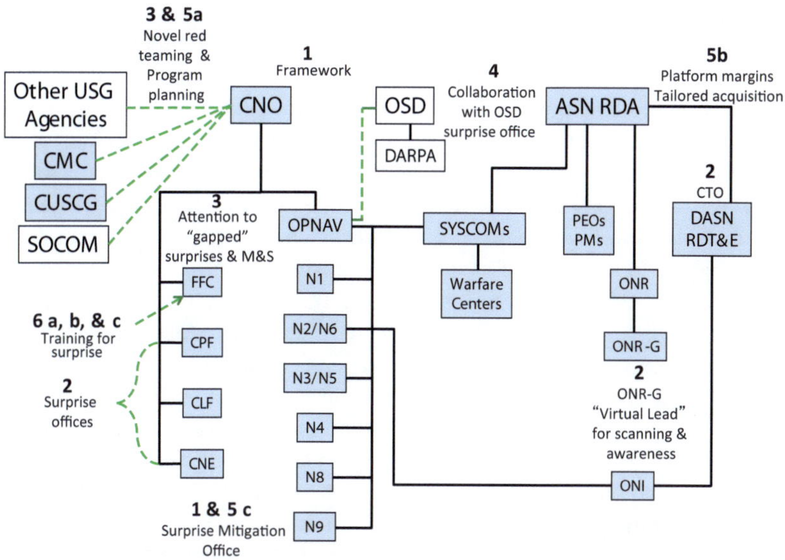

FIGURE 8-3 Mapping of recommendations to organizations within and outside the U.S. Navy, Marine Corps, and Coast Guard. Many of the acronyms have already been identified in the text. See Appendix D for the definitions of those that have not so far been spelled out.

BOX 8-1
Abridged List of Committee Recommendations

1. Implement surprise framework and establish surprise mitigation office to ensure priority capability surprises are addressed and deployed.
2. Combatant commanders and component commands establish surprise mitigation offices with the Deputy Assistant Secretary of the Navy for Research, Development, Testing, and Evaluation (DASN RDT&E) designated as CTO and ONR-G given leadership of scanning and awareness.
3. Initial tasking of surprise mitigation office to focus on horizon scan, modeling, and red teaming involving experts, especially in gapped areas.
4. Collaboration with corresponding Office of the Secretary of Defense (OSD) surprise offices.
5. In planning for surprise,
   (a) Design margins in platforms for future payloads,
   (b) Tailored acquisition, and
   (c) Rapid fielding.
6. In preparing for surprise,
   (a) Realistic planning and training for surprise with collaborating organizations,
   (b) Leveraging of Navy Readiness Reporting Enterprise (NRRE), and
   (c) Broader training for classified capabilities.

recommendations. As shown from the mapping, the organizational impact of each individual recommended change on any part of the organization is rather small. However, the sum of these rather minor changes results in a much more integrated and prioritized organization, in the committee's view.

One area that is not explicitly addressed is the requisite resourcing for surprise phases, especially those where there is no existing program whose resources could be reprioritized. To address the resourcing question requires more detailed consideration of the organizational and process changes that would be needed. However, note that the DOD offices involved in surprise mitigation are resourced for tens of millions of dollars to integrate the early-phase functions.[1] These offices apparently rely on reprioritizations from the respective Services' acquisition and fielding elements to resource development, acquisition, deployment, and training to address surprise. The committee observes that sufficient funds may already exist within the OPNAV, ONR, and ONI organizations to ensure adequate support for the early phases. Reprioritization of acquisition and readiness funds would be required only in the latter phases. If handled within existing programs, internal prioritization, resourcing, and management appears tractable to the committee. However, if the required response to an emerging capability surprise is a new system or major system upgrade, with appropriate program process tailoring, the need and reprogramming might be accomplished within the existing processes and programs in some cases and might require new resources in others.

EXAMPLES OF THE PROPOSED FRAMEWORK FOR THE THREE SCENARIOS

In Appendix A, the three scenarios—(1) loss of space access, (2) social media manipulation, and (3) a natural disaster—are described in detail. In this section, the committee summarizes its understanding of how those activities are being addressed as well as how they could be more effectively addressed with the recommended framework.

(1) Space Access Scenario

How the Space Access Threat Is Being Addressed Today

The committee did not uncover an overarching plan or program to address the general force-level problem of space access denial. Several studies were identified as having provided assessments of vulnerabilities (see Figure 8-4), and certain individual acquisition programs are known to have incorporated antijam designs to ensure operation in projected denied environments. There is

[1] In response to the *Defense Science Board Summer Study on Capability Surprise,* the Office of the Secretary of Defense has established the Strategic Capabilities Office (SCO) in AT&L.

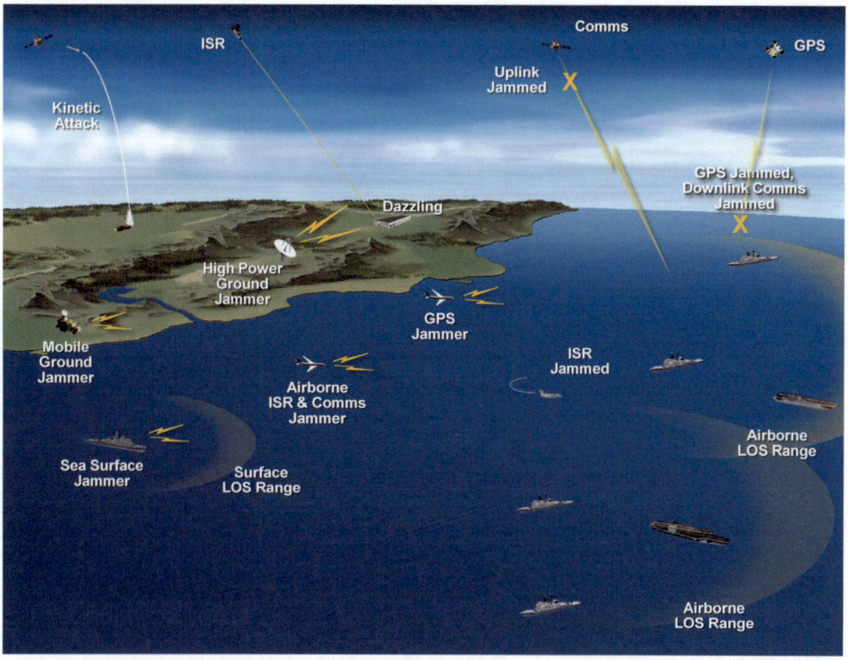

FIGURE 8-4 Space access scenario.

evidence of some consideration for adaptive switching among networks from an individual platform perspective—for example, Space and Naval Warfare Systems Command's (SPAWAR's) Automated Digital Network System (ADNS)—but apparently not from a multiuser network perspective with quality of service needs considered, such as antijam margin, propagation fading, or message error rate. There has been some consideration of how backup services might be provided—for example, via aircraft in lieu of satellites. However, no plans to implement such backup capabilities were identified.

One could expect that perhaps N2/N6 in conjunction with SPAWAR is the appropriate authority to ensure detailed planning for near-term contingencies and longer-term space resilience. However, practical and effective planning would require participation of the platform, weapons, and C2 programs to ensure understanding of impacts and priorities. Further, changes in joint programs such as GPS and certain communications and intelligence, surveillance, and reconnaissance (ISR) systems would likely be limited without substantial funding or DOD direction.

How Proposed Framework Functions Could Mitigate the Space Access Threat

Scanning and Awareness. The threat already appears to have been well characterized by the intelligence community.

Assessing Surprise. It would be expected that a red team would first run models and perform other calculations to determine performance shortfalls of critical individual systems carried on deployed units for the expected geometries. For example, What are the needs for Aegis ballistic missile defense (BMD) cueing by remote radars and Which ISR data are required to plan for prompt strike missions? The resulting inventory of affected systems and impacts could then be used to run a war game series to assess the impact on the campaign objectives and to explore alternative strategies, testing, techniques, and procedures (TTPs) or system modifications to mitigate the impacts.

Prioritization, Option Development, and Decision Formulation. The results and alternative mitigation approaches for the near and far term would be prioritized according to cost-and-time effectiveness. This phase would sponsor the appropriate laboratories and industry to perform the engineering trade-offs to identify potential technical concepts. A trade-off of the concepts would be expected to lead to identification of those few most favorable concepts for practical near-term implementation at reasonable cost and requiring the least complex integration.

Resource and Transition Planning. If prototyping and critical experiments are required, ONR would be expected to lead coordination of that activity, involving laboratories and industry, perhaps in conjunction with SPAWAR. Since joint and other Service programs such as GPS are also likely involved, the surprise mitigation office would coordinate with those program offices, perhaps in concert with appropriate OSD offices. After the prototype testing validates the most expedient approach previously identified via modeling and analyses, a program office would be established or designated to develop the integrated plans and gain industry participation. It is likely that existing industrial players already associated with key elements such as GPS and certain communications and platform capabilities would be selected for limited capability fielding without requiring a time- and cost-consuming competition. However, it may turn out that a more robust, long-term approach is found to be necessary for which a follow-on acquisition program would be appropriate. Then, tailored acquisition processes already in place and determined appropriate to the intended acquisition would be exercised.

Implementation and Fielding. A special project would be established to coordinate the near-term measures such as developing interim airborne alternative services (if that is the selected approach after trade-off analyses and prototype validation), coordinate development of TTPs, and perhaps coordinate longer term developments with both Navy and joint program offices if there is no appropriate program office for longer term acquisition.

Force Response. Fleet Forces Command (FFC) would develop plans for deploying and training for the near-term stopgap measures in conjunction with any joint forces via the COCOMS. FFC would also be involved in ongoing fleet experiments and training to anticipate space access issues and how to respond, thus building resilience and adaptation capabilities among operational users.

(2) Social Media Manipulation

How Such Manipulation Is Being Addressed at Present

The committee learned that some research is under way on how social media like Twitter might be a reliable indicator of population mood as well as on how it could be used by activists, both allied and adversary, as a C2 tool (see Figure 8-5). In addition, studies have been performed on how governments have attempted to thwart such social media, locate leaders via their messages, and influence the crowd by inserting their own messaging and attributions. Recent examples such as the Arab Spring and unrest in London have been studied, for example. This analysis and review is a form of limited horizon scanning. It is also recognized that groups such as special operations forces (SOF) are able to consider the use of social media to promulgate messages and solicit cooperation.

It is believed that the sum total of these research and evaluation efforts would engender expertise in the use of social media to impact crowd behavior, so that if the scenario described in Appendix A about the fictitious place called Provencia were to transpire, experts could be brought into a situation room conference to plan responses. However, except possibly for SOF, there does not appear to be any activity, especially across appropriate U.S. departments, to consider how to strategically leverage social media or respond to threats against U.S. interests facilitated by such media. Also, there does not appear to be significant training or TTP development to prepare forces to respond and prevail in a social media environment.

Using the Surprise Framework to Mitigate Manipulation of Social Media

Scanning and Awareness. Additional effort would be expended to gather intelligence on how others are using, plan to use, and have used social media for crowd manipulation and control. Also, a technology scan of new commercial media products and their projected influence on large numbers of people could be undertaken. If it is determined that evolving social media products present a sufficient threat and/or opportunity over an identified time frame, red teaming be considered.

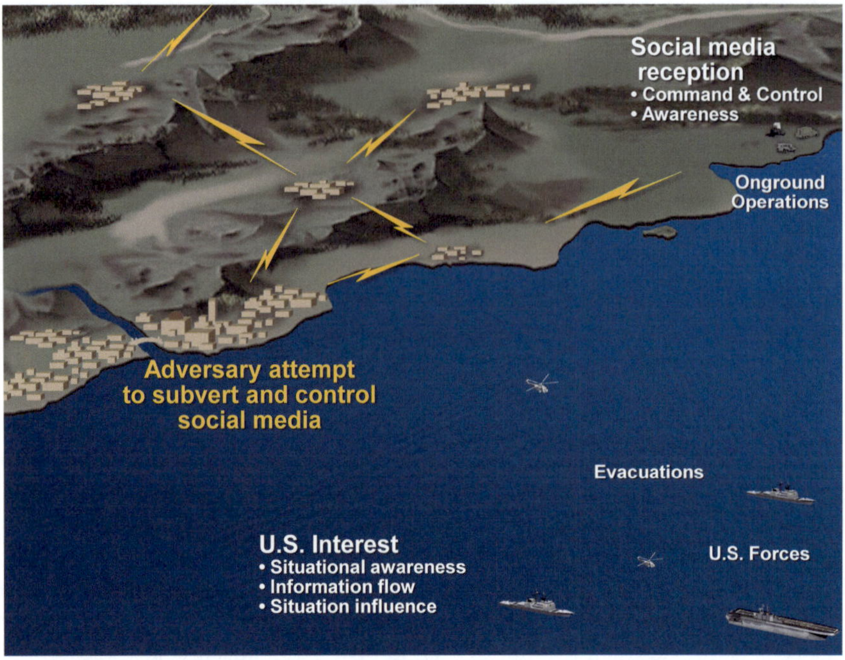

FIGURE 8-5 Social media manipulation.

Assessing Surprise. For this scenario, it is of particular importance to understand the socio-cultural context of trends and unrest around the world that could impact U.S. interests and thus draw in naval forces. To adequately red team potential developments it would be critical to include experts in specific cultures where such scenarios could evolve. Also important would be access to experts with military and political knowledge of the countries or entities of interest. Red teaming would be expected to produce situations where social media could foment particularly volatile protests against U.S. interests. A question then would be, How might U.S. forces respond, including using social media? This, in turn, could lead to identification of which and how many social media assets should be available to naval forces from the command level to the troops. Finally, the red team results could be a backdrop for defining training and exercise needs.

Prioritization, Option Development, and Decision Formulation. This function would lead to definition of potential counters using existing publicly available social media in concert with timely situation awareness knowledge from in-field intelligence. It should also lead to computer modeling of how to detect and interpret the potential intent of adversaries and third parties to manipulate the population. New approaches to situation awareness and command and control may be required. The identified options may not allow numerical trade-off analysis, and as a result, additional red teaming and wargaming may be needed to complement the limited amount of crowd-sourced data for modeling based on human dynamics.

Resource and Transition Planning. One solution might be to resource broader SOF contingency planning, TTP preparation, and more equipping and training of naval forces. It is likely that only limited acquisitions may be needed for the introduction of contingency capabilities. However, over the long term ONR might want to increase research on better predictive and response assessment tools for social media mitigation and leveraging.

Implementation and Fielding. It may be that a limited supply of social media devices and applications should be acquired for the training of key players and for contingency planning. Because of the expected rapid changes in such products, it would be important to plan for limited buys, quick access if needed, and rapid product refresh. Also, as mentioned above, some aspects of social media might begin to be integrated into naval C2 facilities. Whatever decisions are made, it is important that resilience to cyberattacks should be considered, especially since open source products are involved.

Force Response. As actual social media events occur, training and contingency planning might be incorporated into real-world exercises and methodical lessons-learned feedback as actual social media events occur.

(3) Fukishima Disaster

How the Situation Was Addressed from the U.S. Perspective

In Operation Tomodachi, the U.S. Pacific Command (PACOM) was engaged within 24 hours of the earthquake and tsunami in a massive human assistance/disaster relief (HA/DR) operation that involved major contributions from the U.S. Navy, the Marine Corps, the Air Force, and the Army. Operation Tomodachi involved about 24,000 U.S. personnel, 189 aircraft, and 24 naval vessels, with the USS *Ronald Reagan* carrier strike group (CSG) leading the way.[2]

> U.S. efforts focused heavily on transport of relief supplies, SDF personnel [(Japanese) Self Defense Force], and equipment; surveillance of the affected area to search for stranded victims; and restoration of critical infrastructure, such as damaged airfields, in order to sustain operations.[3]

[2]Andrew Feickert and Emma Chanlett-Avery. 2011. *Japan 2011 Earthquake: U.S. Department of Defense (DOD) Response*, Congressional Research Service, Washington, D.C., June 2.

[3]Ibid, p. 4.

The ability to conduct these sorts of operations relied heavily on training within PACOM, on the significant U.S. armed forces presence already resident in Japan (e.g., Yokosuka Naval Base), and on the coordination and training already existing between U.S. and Japanese forces. In fact, Operation Tomodachi was the first time that Japanese helicopters operated from U.S. aircraft carriers and was the first time that U.S. military units operated under Japanese command in actual operations. Key to U.S. participation were contributions to the opening of ports and airfields for supply movement.[4]

U.S. naval activities centered around the use of the USS *Ronald Reagan* CSG to facilitate air operations. USMC operations centered around hands-on ground services to clear transport points, establish relief hubs, and distribute supplies. Air activities concentrated on supply transport and airfield augmentation (e.g., to assist operations at Tokyo's Narita Airport) and included the use of Global Hawk to survey the landscape. Ground activities included the use of U.S. Army personnel already assigned to Japan, including U.S. Army Corps of Engineers personnel.[5]

Lessons Learned from Operation Tomodachi[6]

The official lessons learned are listed below to (1) indicate the areas that could perhaps have been anticipated had there been broader discussions among international HA/DR parties and (2) serve as a guide for what should be routinely done in the future.

- Prescribe HA/DR material to be on hand for future use.
- Improve interoperability training between Japanese and U.S. aircrews.
- Adhere to a memorandum of understanding between U.S. and Japanese forces.
- Provide Voice Over IP (VOIP) capability for CENTRIXS and SIPRNET for deploying strike groups.
- Prescribe and approve (by Commander, Seventh Fleet) a list of standard materials that are carried onboard Commander, Task Force 73's Combat Logistics Force (CLF) assets in addition to the HA/DR kits that could be used in future HA/DR evolutions.
- Create two logistics cells: a tactical cell and a strategic cell.
- Continue to use social media as the number one method to quickly disseminate information to the public at large. Additionally, authorize and fund programs that increase the bandwidth of ships with organic public affairs support.
- Joint Task Force Joint Interface Control Officer needs to push out the correlated common operational picture.

---

[4]Ibid.

[5]Ibid.

[6]U.S.Pacific Fleet, Warfighting Assessment and Readiness Directorate, personal communication with NRC staff officer, July 27, 2012.

- Employ F/A-18 Shared Reconnaissance Pod (SHARP) capability in HA/DR when available. Institute capability doctrinally.
- Formalize a method for coordinating the transportation of nongovernmental organizations (NGOs) or civilian-provided HA/DR supplies to reduce confusion about the process during the next HA/DR mission.
- Ensure that all concerned are aware of the information-sharing agreements and systems.

The Navy clearly had a preparation and readiness capability learned from previous disasters and from ongoing procedures and contingencies for nuclear systems safety. It is important to note that the U.S. Navy has decades of experience operating nuclear-powered ships and submarines and in protecting personnel in a chemical, biological, radiation, or nuclear (CBRN) environment.

Particularly important, the Navy had the vast experience of HA/DR associated with Operation Unified Assistance, which was established to help with the massive Indian Ocean earthquake and resulting tsunami on December 26, 2004. This disaster, the deadliest tsunami on record (approximately 230,000 dead in 14 countries),[7] served as excellent training in the use of naval forces for HA/DR (the USS *Lincoln* Battle Group), and cooperation and coordination with the military resources of Australia, Japan, Singapore, Russia, France, and Malaysia.[8]

Specific Lessons Learned from Operation Unified Assistance

- Importance of tsunami early warning systems and damage reporting. "An Indian Ocean tsunami early warning system could have saved many lives."[9] Also, had more accurate damage reporting been available, the U.S. Navy and other disaster relief assets could have begun their operations earlier.[10]
- Importance of rapid and flexible sea basing. Sea basing is essential for disaster areas with little or no infrastructure. Operation Unified Assistance's sea basing served as a model for post-Hurricane Katrina cleanup.[11]

---

[7]U.S. Geological Survey. 2013. "Magnitude 9.1 – Off the West Coast of Northern Sumatra, Earthquake Summary," Significant Earthquake Archive, online, last modified February 15. Available at http://earthquake.usgs.gov/earthquakes/eqinthenews/2004/us2004slav/#summary. Accessed February 27, 2013.

[8]Bruce A. Elleman. 2007. *Waves of Hope: The U.S. Navy's Response to the Tsunami in Northern Indonesia*, Newport Paper 28, Naval War College Press, Newport, R.I., February. Available at http://www.usnwc.edu/Publications/Naval-War-College-Press/Newport-Papers/Documents/28-pdf.aspx. Accessed February 13, 2013.

[9]Ibid, p. 90.

[10]Ibid, p. 90.

[11]Ibid, p. 91.

Within the Surprise Framework

Although considerable naval experience was brought to bear, and multiple agencies and departments were involved as well as international partners, there appear to be a number of additional steps to be taken using the proposed framework. In the case of Tomodachi, therefore, the committee recognizes that the key elements of the operation were already operable. However, in the interest of disclosing "best practices," the committee notes that, in particular, the Coast Guard and Marine Corps have contingency units that continually operate all the elements of the framework. Therefore, from the standpoint of the surprise framework, the committee suggests what additional measures might be considered.

Scanning and Awareness. A team could explore how emerging technologies, such as more interoperable radios, portable cell towers, and portable power might be used to better prepare for future incidents. Assessing lessons learned from the introduction of new experimental technologies as events unfold could also inform the decision on adding new capabilities. For example, during the Haiti disaster, airborne imaging sensors were brought in from laboratories such as the Massachusetts Institute of Technology Lincoln Laboratory (MIT/LL) and the Johns Hopkins University Applied Physics Laboratory (JHU/APL) to track population movements for better coordination of aid. This opportunity revealed the value of the sensors and suggests how they might best be used in future disasters.

Assessing Surprise. The Coast Guard and the Marine Corps appear to merge red teaming with training to prepare for future incidents. This arrangement—merging of red teaming and training—could be extended to all naval forces as well as to other government agencies and perhaps key regional allies and partners.

Prioritization, Option Development, and Decision Formulation. There is a huge body of experience and lessons learned from response to disasters in CONUS and to disasters far from the United States. In some cases there have been political implications, e.g., U.S. assistance to Iranian fishermen, and in other cases, governments have prevented or resisted attempts to conduct rescue and assistance. It also appears that some lessons learned and postevent assessments are not always taken to heart by all parties. It is therefore suggested that some scenarios developed by red teaming be considered for multiservice/department contingency planning. As part of the planning the requisite assets and their prepositioning implications could be determined. Further, different approaches to accountability, both within the United States and with other countries, should be anticipated and accommodated in the planning. This activity could lead to several capability "packages" that resemble contingencies that the Marine Corps and Coast Guard have developed, but on a larger scale.

Resources and Transition Planning. Lessons learned and technology introduction opportunities can inform prioritization of assets, interdivisional policies and practices, and planning by expected responders.

Development and Implementation/Force Response. Global emergencies that occur from time to time have resulted in continuing resourcing, replenishment, contingency planning, and training opportunities for the Coast Guard and the Marine Corps, broadening their naval scope.

THE WAY AHEAD

All naval forces of the world have been nurtured in an environment that depends heavily on an individual's ability to deal with surprise. A professional mariner was often judged on his ability to "read" the wind and the sea or to "weather" a storm without loss of limb or loss of the vessel. This single-handed ability of the captain of a vessel to deal with the surprises is a classic template that has colored naval operations from the beginning.

Thankfully, tremendous advances in technology and information sharing have given ship captains enhanced tools and data with which to face today's surprises—as long as the event or one similar to it has been previously experienced and a solution to it has been documented.

However, when a totally new surprise emerges, it takes strong leadership to steer away from "let the captain handle it" or "let the commander and his staff figure this one out." An ad hoc approach to facing a new problem is not likely to result in a high-quality solution and is even less likely to be worthy of attribution to the mature and capable naval forces of the United States.

Historically, the Navy, the Marine Corps, and the Coast Guard have registered some remarkable successes based on timely and thoughtful ad hoc reaction to surprise. Similarly, some solutions have been less than stellar.

The goal of naval forces must be to always find the best reaction to a surprise, using the fullest measure of knowledge, intelligence, experience, and talent that can be brought to bear.

Appendixes

A

Scenarios

SCENARIO 1: DENIED ACCESS TO SPACE—AN INTELLIGENCE-INFERRED SURPRISE

It has been widely reported that several key U.S. space capabilities would be vulnerable to loss or disruption in an antiaccess/access denial (A2/AD) situation. The primary capabilities lost are (1) access to timing and position via GPS, (2) access to communications via commercial or military communications satellites, and (3) reception of ISR signals and imagery. Although our naval forces could surely inflict similar disruptions on an adversary's space systems, the impact of potential loss of significant access by friendly forces, including those of the United States, is not diminished. The following sections describe the potential attack modes.

Electronic Attack

The most common and likely threat is electronic attack. The proliferation of military and commercial radio frequency (RF) transmitters makes the ability to interfere with satellite receivers a matter of geometry, transmitted power level, and selection of effective waveforms and operating bands. If provided with sufficient jamming power and appropriate waveform selection, then space-based communications, ISR, and GPS systems could be affected if within line of sight of an adversary.

It has long been known that GPS receivers are relatively vulnerable to interference. A number of U.S. weapon systems programs have anticipated potential GPS vulnerabilities in their designs. Accordingly, they may provide backup navigation and timing capabilities or antijam design features to afford additional antijam margin. On the other hand, some systems do not yet have such features. Further, to the best of the committee's knowledge, the combined impact of substantial GPS jamming from multiple sources against naval forces has not been well characterized but is expected to be significant.

Electronic jamming of communications satellites by interfering with received communications signals at either the relaying satellites or at the airborne or surface vehicles themselves can also be highly effective, depending on the antijam protection level of the communications networks. It would be expected that the lower frequency systems with less antenna directivity would be more vulnerable because there would be greater geometric opportunity to enter the receive antenna beams. Commercial systems, often used by the military, are typically not built to operate against overt electronic jamming and would therefore be expected to be relatively vulnerable to a determined electronic attack. The present diversity of available communications satellite networks can substantially complicate an adversary's electronic warfare (EW) attack plan. On the other hand, attempting to adapt to changing network availabilities during an electronic attack can complicate friendly force operations, especially with the lines of communications being interrupted. In particular, interoperability is highly volatile in such cases because some units may be without access to some networks while others are without a different set of networks depending on jammer-victim geometries. Thus an attack on only a subset of network terminals could significantly impact the entire network by limiting the commonality of a unit's situation awareness. As far as the committee knows, this integrated effect and identification of countermeasures to facilitate adaptation have not yet been examined.

The ability to induce interference, loss of sensitivity, and perhaps deceptive signal inputs to ISR sensors has also been discussed for many years. The diversity of sensors and their locations can complicate planning and operations by adversaries to jam communications networks. However, the ability to not only impact ISR asset availability but also degrade the credibility of their sensor data by means of deceptive waveforms can result in unreliable, untrustworthy information.

Kinetic Attack

It is well known that countries such as China and Russia possess antisatellite capabilities. Other countries are seeking such capability as well. Even if a country initially only possesses the ability to destroy satellites in low Earth orbit (LEO), the ability will probably extend to higher orbits in just a matter of years. It is further recognized that a mutually assured destruction standoff could occur in which both sides—the United States and an adversary—are capable of attack. However, the adversary may determine that attacking U.S. assets could have a greater negative effect on its units deployed far from U.S. territory than loss of overhead access at the country denying entry, which can use land- and air-based alternatives more effectively for geographic advantage.

Cyberattack

A cyberattack can have multiple entry points and can affect far more than just space assets. But it may be especially advantageous for an adversary to impact naval command and control via communications, GPS, and ISR early in an operational escalation in order to deny us knowledge of our adversary's movements, or, conversely, to deny them knowledge of our movements.

Sequence of Events

This scenario is illustrated in Chapter 8, in the section "Examples of the Proposed Framework for the Three Scenarios" (the figure is shown here again as Figure A-1).

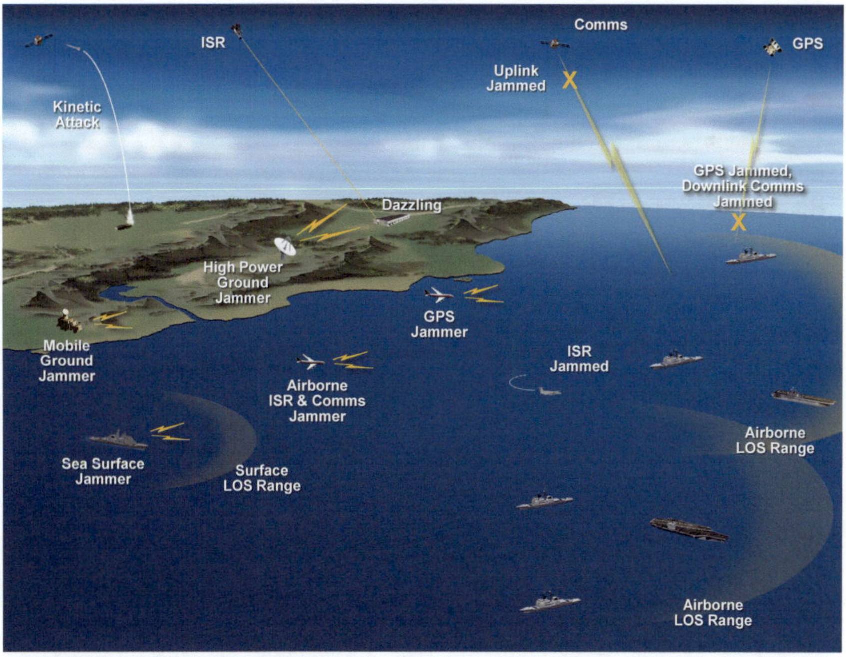

FIGURE A-1 Space access scenario.

1. As a provocation occurs and U.S. forces converge toward a denied area, the adversary country initiates a series of coordinated electromagnetic attacks on our communications, GPS, and ISR using land-based and airborne jammers. First, the attacks are intermittent, as if to serve as warnings. They become more frequent and sustained as the adversary learns how to operate and observes U.S. reactions.

2. Those U.S. assets most vulnerable and/or closest to the jamming sources are disabled immediately, and our naval forces have only limited access to situation awareness data from the front. Further, reception of command direction from the Missions Operations Center is denied except for those forces with priority antijam circuits and systems.

3. U.S. forces then begin to attack the adversary's space systems. However, because the theater of operation is near the adversary's borders and the adversary has other land-based and short-range communications alternatives, the impact of the U.S. attack is contained.

4. As part of the escalation the adversary opens up the attack to include cyber operations on military nodes and capital units.

5. An effort is made to limit the traffic on priority networks to the capital units with the expectation that local tactical networks such as Link 16 can be used to transmit information to and from the noncapital units via tactical links for those units for which link jammers are beyond their horizons. Whereas some weapon systems and platforms are operable in GPS jamming because they have alternative capabilities or antijam margins, other units and weapons without such features are limited in their ability to deliver precision strikes.

6. U.S. forces must respond and coordinate their activities in an ad hoc manner without having the communications connectivity or confidence in situational awareness data to ensure appropriate counteractions and access to transmitted orders. This deficit results in degradation of coordinated strategic and tactical actions.

SCENARIO 2: CROWDSOURCING VIA SOCIAL MEDIA—A DISRUPTIVE SURPRISE

This scenario is illustrated in Chapter 8, in the section "Examples of the Proposed Framework for the Three Scenarios" and includes the following components:

- Incorporate USN, USMC, USCG,
- Determine overseas location,
- Exploit media for world opinion, create doubt, and rally discontent aimed at the United States, and
- Use social media to interfere with the execution of naval missions.

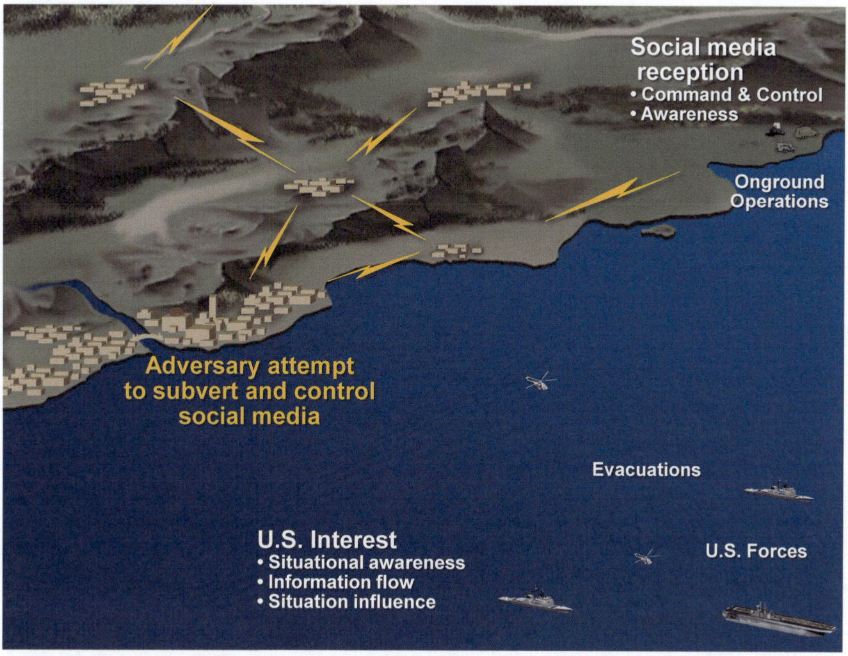

FIGURE A-2 Social media manipulation.

To explore the topic of social media in this context, the committee posited a scenario taking place in a country called Provencia. Tensions were high in Provencia. The Freedom for Provencia (FFP) (shadow organization for the international drug cartel) was making great progress in unseating the legitimate government of the country. With its comparatively large disposable incomes, generous bribes, ruthless actions, false promises, and cunning nature, FFP initiated the final stages of its campaign, to replace the country's government and to expel all foreign law enforcement and military personnel.

The FFP's main job was to clear the area and keep the USCG *Gallatin* (WHEC 721) away from it, for she had interdicted their last two drug shipments. An opportunity presented itself when the *Gallatin* became entangled with some abandoned fishing nets during a night mission. The action was captured on film and spliced in with film of a local fishing boat collision. This modified film clip showed a collision in which both fishing boats were lost along with several crew lives. The doctored film was sensationalized and released on YouTube and several other social networks. Industry and local populations were outraged. Media activists were recruited to magnify the problem and create additional "documented" transgressions. The following week another staged night accident involving the *Gallatin* occurred. Again, the media activists exploited the "arrogant" actions of the United States. The issue was introduced to the U.N. Fisheries Panel. Pressure was mounting for the *Gallatin* to leave the area and to conduct only daylight operations to lessen the chance of continued "accidents."

Ashore in the jungles of Provencia similar actions were planned for discrediting the U.S. presence in the country. The intentional night spraying of a village with noxious herbicide by helicopters caused havoc and several deaths among villagers. As the Americans from the USS *New York* (LPD-21) were thought to have the only helicopters, the United States was blamed and local activists provided stories and pictures to all the media outlets. A medical team dispatched from the USS *New York* was ambushed en route to the village and forced over a cliff and into a gorge with the loss of U.S. lives. A dam at the bottom of the gorge near the accident site was destroyed by the cartel, resulting in a flood downstream with more loss of lives and property. Explicit raw footage of the area was played on social networks and identified by the world media as the work of U.S. agents.

Village personnel traveling to an NGO medical center were attacked by personnel wearing U.S. military uniforms who spoke English. The NGOs at the medical center, representatives from several countries, took statements from the wounded villagers and filed stories with their media sources. Two American citizens visiting Provencia's waterfront capital were beaten and robbed, and several other attacks on Americans occurred that evening in the night life part of town. The Marine security guardhouse located near the American counsel was firebombed. Marines rushing to their burning house became involved in an auto accident. All these incidents were planned, executed, and filmed by FFP operatives. Then, not only the social networks but also radio and television networks carried and expanded the stories, generating huge controversy about the U.S. presence in Provencia. A bomb went off in the market square, and personnel in American military uniforms were seen leaving the area. An American employee at the U.S. consulate and his family were attacked while attending a school soccer game, and a bonded U.S. warehouse was broken into, looted, and set afire. Crowds formed around the U.S. facility and demanded that the Americans leave their country.

The government and local law enforcement were incapable of controlling the protestors, and riots were orchestrated by the FFP. The U.S. consul general ordered the evacuation of all nonessential personnel and requested additional security forces. As evacuees gathered at embarkation points, the FFP, using the social media, called on flash mobs to inundate the embarkation sites and block all movement. The USS *New York* flew in a platoon of Marines to protect the consul; however, all planned landing zones were blocked by flash mobs. The USS *New York* dispatched landing craft toward the harbor entrance to initiate a surface evacuation option, but the entrance and channel were blocked by numerous anchored vessels.

U.S. forces finally reached the U.S. consul early the next morning after a clandestine night raid by the SEALs from the USS *New York* cut the anchor lines to several of the blocking vessels. The circling surface craft flowed through the breach, landed on the beach, and discharged troops, who made their way rapidly to the consulate. Shortly thereafter a landing zone was secured and follow-on forces arrived to restore order and security at the consulate and complete the mission. See Figure A-2 for an illustration of this scenario.

SCENARIO 3: NATURAL DISASTER—A DISRUPTIVE SCENARIO

This scenario is based on a recent disaster, the Japanese earthquake and tsunami on March 11, 2011, and the resultant major nuclear accident at the Fukushima Daiichi nuclear power plant. The following events came into play:

- Largest earthquake to ever hit Japan,
- Among the five most powerful earthquakes ever recorded,
- Resulting tsunami wave reached a height of 40 meters, with a 14- to 15-m wave height in the vicinity of the Fukushima Daiichi power plant,
- More than 22,000 people died or went missing,
- Major accident was set off at the Fukushima Daiichi nuclear power plant,
- Earthquake was "surprising" because it had not been anticipated by many seismologists,[1] and
- U.S. DOD involvement, known as Operation Tomodachi, was major.

The Fukushima Daiichi power plant suffered damage to nuclear fuel, the reactor pressure vessels, and the primary containment vessels of multiple reactors. There was also a large release of radioactivity, about one-sixth of that associated with Chernobyl.[2] For reference, the Fukushima Daiichi power plant was designed to withstand an 8.2 earthquake and a 5.7 meter tsunami.[3]

The report of independent assessment of the events associated with the Fukushima Daiichi power plant accident is now public.[4] The key finding from this independent report is that the accident was "a profoundly manmade disaster—

---

[1]According to James Mori of the Disaster Prevention Research Institute, Kyoto University, at the 2012 Annual Meeting of the American Association for the Advancement of Science, Vancouver, British Columbia. See Clara Moskowitz, 2012, "What We Learned from Japan's Deadly Earthquake: One Year Later," *OurAmazingPlanet*, online, February 21. Available at http://www.ouramazingplanet.com/2479-japan-tohoku-earthquake-lessons-learned.html. Accessed February 13, 2013.

[2]Fukushima Nuclear Accident Independent Investigation Commission. 2012. *The Official Report of the Fukushima Nuclear Accident Independent Investigation Commission, Executive Summary*, The National Diet of Japan, Tokyo. Available at http://warp.da.ndl.go.jp/info:ndljp/pid/3856371/naiic.go.jp/en. Accessed February 13, 2013.

[3]J. Buongiorno, R. Ballinger, M. Driscoll, et al. 2011. *Technical Lessons Learned from the Fukushima Daiichi Accident and Possible Corrective Actions for the Nuclear Industry: An Initial Evaluation,* MIT-NSP-TR-025 Rev. 1, Center for Advanced Nuclear Energy Systems, Massachusetts Institute of Technology, Cambridge, Mass., July 26. Available at http://web.mit.edu/nse/pdf/news/2011/fukushima-lessons-learned-mit-nsp-025_rev1.pdf. Accessed February 28, 2013.

[4]Fukushima Accident Commission. 2012. *Official Report, Executive Summary*, The National Diet of Japan, Tokyo. Available at http://warp.da.ndl.go.jp/info:ndljp/pid/3856371/naiic.go.jp/en. Accessed February 13, 2013.

that could and should have been foreseen and prevented."[5] The report further concludes that the plant had insufficient training and that the accident "was the result of collusion between the [Japanese] government, the regulators and TEPCO [Tokyo Electric Power Company]."[6] Clearly, a combination of conflicts of interest and inattention to contingency planning and safety procedures significantly contributed to the incident. Although heroism and public cooperation were much extolled at the time, the lack of material, contingencies, and C2 capability were later identified as contributing to the scale of the disaster—in particular, to the nuclear meltdown.[7]

Several key lessons learned from the Fukushima Daiichi nuclear power plant accident include the following:

- TEPCO was too quick to cite the tsunami as the cause of the nuclear accident and to deny that the earthquake had caused any damage.
- The power plant situation continued to deteriorate because the crisis management systems of the Kantei (Prime Minister's Office), the regulators, and other responsible agencies did not function properly.
- Residents' confusion over the evacuation stemmed from the regulator's negligence and failure over the years to implement adequate measures against nuclear disaster.
- Planning to prepare for unanticipated risks was lacking.
- The supply of power for emergency use at the power plant needs to be enhanced, and the vulnerabilities of the power supply system need to be remediated.[8]

[5]Ibid, p. 9.
[6]Ibid, p. 16.
[7]Ibid.
[8]Ibid.

B

Exemplars

OPERATION BURNT FROST

The committee reviewed the events leading to rapid modification of the Aegis ballistic missile defense (BMD) combat system and associated national systems that led to the shooting down of a wayward National Reconnaissance Office (NRO) satellite within only a few weeks of an analysis and then a decision by the Bush administration. It appears to the committee that the Aegis BMD and other national programs have the key elements of their functional frameworks and organizational constructs embedded in the respective program infrastructures that enable quick reaction to a disruptive surprise.

Within the Framework

There were some lessons learned, which are as follows:

Organization. The R&D, acquisition, implementation, and operational authority chains within Aegis BMD, in collaboration with the Missile Defense Agency (MDA), are mature and based on the long-established Aegis program. The NRO authority chain provided information on the satellite configuration and status as well as updates on its orbit and tumbling movement for targeting purposes. However, since the capability for Aegis antisatellite operation was not in place, some authority discontinuities and operational connectivity issues had to be addressed. Separate security processes required special attention to gain release of certain information.

Scanning and Awareness. This function is not likely to have pointed to the need to prepare for such an action in the name of national safety. Even if it had, the multiplicity of variables in such a scenario would have made it impossible to prepare. As it happened, when the request came to determine if such a capability could be rapidly implemented, the existing expertise and the associated physics and systems models and databases allowed a quick answer.

Assessing Surprise. A vigorous debate was enabled among experts from the laboratories, government, and industry that allowed realistic consideration of the risks and identification of needed information.

Prioritization and Decisions. Given the emergency nature of the situation and the high-level decision makers it had attracted, the "crash effort" was expedited.

Implementation and Fielding. Because Aegis BMD is a highly organized acquisition program, modifications could be made rapidly, and procedures for verification, testing, and certification could be expedited because the maturity of the program, technical expertise, and special facilities enabled informed expert tailoring of baseline modifications and procedures. Further, the changes were primarily in software, allowing relatively rapid changes to be implemented and recertified via system-in-the loop facilities and test beds.

Force Response. Rapid development and rehearsal of procedures from the command to the crew of the USS *Lake Erie* were required. Technical assistance from the laboratory and the contractor community was supplied, as needed, at every level.

Lessons Learned

Operation Burnt Frost involved the authorized shooting down of a wayward NRO satellite. A number of lessons learned surfaced in connection with the following:

- Procedural problems,
- Operational connectivity—members, expertise, subject matter expert (SME) level, etc.
- Authority discontinuities—standard operating procedures (SOPs) for chains of command, priority assessment, and
- Special sensitive information authority (security experience, risk-actual, and security).

The difficulties encountered with this event can be mitigated by adopting the concept prepared by a former Commandant of the Coast Guard, ADM Thad

Allen—namely, by identifying possible contingencies and developing standing mission response teams. First, the specific likely contingency situations or conditions are developed. After examining the contingency excursion, SMEs and informed experts from appropriate discipline fields are identified as team members for the scenario or contingency. These individuals are assigned to a contingency team. They gather for meetings and briefings prior to an actual event and are on on call and activated when an actual event occurs. Channels of communication, protocols, and relationships are worked out prior to activation. Participation and membership can be at the level of an action officer, a deputy, or a principal.

AIR FORCE RED TEAM

The Air Force has operated a red team program, the Air Vehicle Survivability Evaluation Program, for many years, with technical leadership by the Lincoln Laboratory at the Massachusetts Institute of Technology. The program focuses on the threats to tactical air penetration of an adversary's defenses for effective precision engagements. The approach covers all the elements of the framework as follows:

Scanning and Awareness. Technical intelligence is continually collected, and follow-up information requests are made as needed. From this information a projected threat is modeled in terms of its technical characteristics, the projected time frame, and how effectively it performs against U.S. systems.

Assessing Surprise. A structured, scenario-based approach allows assessing the impact of the emerging threats to a mission and evaluating potential material and nonmaterial approaches to mitigating the threat, through use of modeling and simulation.

Prioritization and Decisions. The results are evaluated and program revisions are made as needed, whether the mitigation is by development of new tactics, techniques, and procedures (TTPs), modifications to an existing system, or start of a new acquisition.

Implementation and Fielding. For material solutions the standard acquisition processes are followed. If a quick reaction is needed, the program is capable of adapting in a manner similar to Aegis BMD.

Force Response. The program includes testing, force introduction, training, and assessment activities associated with acquisition programs. The advantage of this program approach is that it provides authority to ensure resourcing and collaboration for those solutions requiring changes to multiple, interoperating systems.

SSBN SECURITY PROGRAM

The SSBN Security program began in the 1970s to examine, through theory and experiment, any potential phenomenon or technology that might enable detection of deployed nuclear-powered ballistic missile submarines (SSBNs). Over the years it became clear that the same approach could be extended to attack submarines (nuclear) SSNs. Today the program continues to carry out security against the detection of SSNs, SSBNs, and nuclear-powered, guided-missile submarines (SSGNs).

Scanning and Awareness. The program relies not only on technical intelligence but also on the monitoring of scientific communities by technical experts in key areas for identification of any phenomena and technologies that might be of interest. Those are prioritized in terms of technical risk and risk to the missions.

Assessing Surprise. A review process allows objective consideration of the potential for detection and under which conditions. Modeling and scientific expertise are coupled with deep operational knowledge and prior at-sea experimental results. Recommendations are provided on (1) what bears watching for further development and (2) what needs further prioritization in light of the development stage—for example, proof-of-concept demonstration or gathering of experimental data.

Prioritization and Decisions. In accordance with recommendations, plans are developed with resource allocations against the funding line.

Implementation and Fielding. Scientific investigations and prototyping are conducted by national and naval laboratories and naval activities in coordination with the fleet. If results indicate that design changes are needed or new capabilities must be acquired, plans are developed with industry for implementation.

Force Response. Sea trials and training are conducted as required.

C

Biographies of Committee Members and Staff

Jerry A. Krill is assistant director for science and technology and chief technology officer at the Johns Hopkins University (JHU) Applied Physics Laboratory (APL), leading APL's innovation initiative and establishing a new Research and Exploratory Development Department. Previously he served as the JHU/APL assistant director for programs and chief quality officer. In that position Dr. Krill was responsible for all of APL's more than 700 programs, implemented an ISO-based quality management system, and co-chaired milestone and program management reviews for APL's NASA science missions and instruments. Earlier positions at JHU/APL included executive for air defense programs and head of the Power Projection Systems Department with its precision engagement and info-centric operations program portfolios. He holds a doctorate in electrical engineering from the University of Maryland. Dr. Krill's expertise includes combat systems, systems engineering, sensor and weapons networks, and microwave technology. He was a principal in developing the U.S. Navy's Cooperative Engagement Capability, which networks air defense systems, and, in 2000, led a joint Navy/Ballistic Missile Defense Organization (BMDO) working group to develop technical concepts for the Navy's role in national and regional missile defense. He holds 18 patents, and his awards include Innovator of the Year by the *Baltimore Daily Record* and the American Society of Naval Engineers "Jimmie" Hamilton Award. A former member of the Naval Studies Board (NSB), he has served on NRC and Defense Science Board (DSB) studies. Other memberships include the Institute of Electrical and Electronics Engineers (IEEE), American Institute of Aeronautics and Astronautics (AIAA), National Defense Industrial Association (NDIA), National Space Society (NSS), and the International Council on Systems Engineering (INCOSE).

J. Paul Reason, ADM, USN (Retired), is currently an independent consultant, having retired from the U.S. Navy with the rank of admiral after 35 years of service. His background includes naval and joint operations, as well as Department of Defense planning, programming, and budgeting. In his last position, he served as Commander-in-Chief, U.S. Atlantic Fleet, where his responsibilities included the training, maintenance, and readiness of naval forces deployed to the Mediterranean and Caribbean seas, South America, and the Persian Gulf. He was also responsible for the operations of most U.S. Navy bases and facilities along the East and Gulf coasts of the United States, in Puerto Rico, Cuba, and Iceland. He has served on numerous scientific boards and advisory committees, including as a member of the NRC Committee on the "1,000-ship Navy"—A Distributed and Global Maritime Network and the Committee on U.S. Forces' Capabilities for Responding to Small Vessel Threats; he is a member of the Naval Studies Board.

Ann N. Campbell is currently director for information solutions and services at Sandia National Laboratories. Her organization develops and stewards Sandia's enterprise applications and information environment, develops custom software applications to support national security missions, and performs cybersecurity research. Dr. Campbell was previously senior manager and deputy for cyber research in Sandia's Chief Technology Office, with responsibility for development and implementation of an institutional strategy for cyberresearch. She has also served as acting director for Sandia's Cyber Security Strategic Thrust and as senior manager deputy for technical programs for the laboratory's Defense Systems and Assessments strategic management unit. As senior manager for assessment technologies in the Information Systems and Analysis Center at Sandia, Dr. Campbell led the development and management of a portfolio of programs focused on vulnerability assessments and development of national security solutions for multiple government sponsors. She has served on several National Research Council committees focused on technology surprise, including TIGER and the Committee on Science and Technology for Defense Warning. Dr. Campbell holds a B.S. in materials engineering from Rensselaer Polytechnic Institute and an S.M. and a Ph.D. in applied physics from Harvard University.

Timothy P. Coffey is an independent consultant having recently retired as the Edison Chair at the Center for Technology and National Security Policy at the National Defense University. He graduated from the Massachusetts Institute of Technology (MIT) in 1962 with a B.S. in electrical engineering and obtained an M.S. (1963) and a Ph.D. (1967), both in physics, from the University of Michigan. Dr. Coffey joined the Naval Research Laboratory (NRL) in 1971 as head of the plasma dynamics branch, Plasma Physics Division. In this position, he directed research in the simulation of plasma instabilities, the development of multidimensional fluid and magnetohydrodynamic codes, and the development of computer codes for treating chemically reactive flows. In 1975, Dr. Coffey was named superintendent, Plasma Physics Division; he was appointed associate director of research for general science and technology in 1980. Two years later, Dr. Coffey was named director of research at NRL. Today, he serves on numerous scientific boards and advisory committees such as the recently completed NRC Committee on Operational Science and Technology Options for Defeating Improvised Explosive Devices; he is a member of the NSB.

Stirling A. Colgate (NAS) is a physicist at the Los Alamos National Laboratory and a professor emeritus of physics at the New Mexico Institute of Mining and Technology (New Mexico Tech), where he still continues an experiment on the origin of the magnetic fields of the universe. During the Second World War he served in various positions in the U.S. Merchant Marine. After the war, Dr. Colgate returned to Cornell University, completing his B.S. and a Ph.D. in nuclear physics, then taking up a position as a postdoctoral fellow at Berkeley. In 1952, he moved to Livermore National Laboratory, where he performed the diagnostic measurements for the nuclear tests of the hydrogen bomb just developed at Los Alamos National Laboratory. He was responsible for the design and execution of the fast nuclear diagnostics (gamma rays, neutrons, and x-rays) of the Bravo test, 15 megaton equivalent yield, including a dozen vacuum pipelines 2 miles long. Later, he served as the scientific advisor to the State Department during the test ban negotiations in Geneva, where he proposed the mutual need for the detection of nuclear testing in space by use of spy satellites. The surprising Soviet acceptance of this concept predated perestroika. Dr. Colgate went on to serve as president of New Mexico Tech from 1965 to 1974, where he also conducted research in astrophysics and atmospheric physics. He became an adjunct professor at New Mexico Tech, moving to Los Alamos National Laboratory, where he currently continues work in astrophysics and inertial confinement fusion. Dr. Colgate is a fellow of the American Physical Society (APS) and the American Association for the Advancement of Science (AAAS); he is a member of the American Astronomical Society (AMS) and the American Meteorological Society among others; and he was a founding board member of the Santa Fe Institute.

Charles R. Cushing (NAE) is president of C.R. Cushing & Co., Inc., a firm of naval architects, marine engineers, and transportation consultants with offices in New York and Europe. He has been responsible for the design and/or construction of over 250 ocean-going vessels in the United States, Europe, and the Far East. Specifically, he has directed the concept, preliminary, and contract design; strategic planning; plan approval; and supervision of construction of vessels from tankers and container ships to bulk carriers and passenger ships. His work has included new construction, conversion, repair, and refurbishment of vessels. Dr. Cushing has been directly responsible for risk analyses, safety audits, energy audits, and the preparation of the U.S. Coast Guard's *Tankerman's Manual*. He has designed intermodal shipping containers and a myriad of container handling equipment, and he holds a number of patents in maritime and intermodal technology. For 26 years he has taught a course on ship acquisition and for 7 years a course on maritime casualty investigation, both at the United Nations World Maritime University. For 12 years he has served on the final selection committee of the National Shipbuilding Research Program, which sponsors and funds naval and commercial shipbuilding research in the United States. Dr. Cushing has served on scientific boards and advisory committees, and he is currently a member of the NRC's Marine Board as well as a member of the NSB.

Susan Hackwood is currently executive director of the California Council on Science and Technology (CCST) and professor of electrical engineering at the University of California, Riverside, where her research interests include electrical engineering, signal processing, and cellular robotic systems, to name just a few. CCST is a not-for-profit corporation comprised of 150 top science and technology leaders sponsored by the key academic and federal research institutions in California, and it advises the state on all aspects of science and technology including nanotechnology, stem cell research, intellectual property, climate change, energy, information technology, biotechnology, and technical workforce development and education. Dr. Hackwood has worked extensively with industry, academic, and government partnerships to identify policy issues of importance and is active in regional and state economic development. Dr. Hackwood is a fellow of the IEEE and the AAAS. Dr. Hackwood received a Ph.D. in solid state ionics from DeMontfort University. She has served on other scientific boards and advisory committees including membership of the NRC Committee on Improving the Decision Making Abilities of Small Unit Leaders; she is a member of the NSB.

Lee M. Hammarstrom is special assistant to the director at the Applied Research Laboratory of Pennsylvania State University. Previously, he was the first chief scientist at the National Reconnaissance Office (NRO) and chief scientist at the Office of the Secretary of Defense for Command, Control, Communications, and Intelligence. Mr. Hammarstrom has broad expertise in areas ranging from technology development to the testing and deploying of military and intelligence systems. He has served on numerous scientific and advisory committees and he is an NRO Pioneer and senior fellow.

Nathaniel S. Heiner is a Northrop Grumman technical fellow as well as director and principal architect, C4I integration, for Northrop Grumman Corporation's Technology and Engineering Group. He previously worked as the U.S. Coast Guard's senior civilian officer for technology, often acting as the Coast Guard chief information officer. In his prior tour with Northrop Grumman and the Federal Data Corporation in the 1990s, he was director of Web/Internet security services, focusing on emerging threats to Internet-based systems. He spent his early career as a UNIX networking expert, writing networked database applications and peripheral drivers and securing communications systems law firms, Congress, AT&T, and MCI. Specializing in mathematical logic and linguistics, Dr. Heiner earned Ph.D., M.Phil., M.A., and B.A. degrees at Columbia University, where he also taught.

Leon A. Johnson, Brig Gen, USAFR (Retired), is currently an independent consultant having retired from the U.S. Air Force with the rank of brigadier general after 33 years of service. During his career, General Johnson commanded a fighter squadron, fighter group, was the vice commander of 10th Air Force at the Joint Reserve Base in Ft. Worth, Texas, and served as mobilization assistant to the Assistant Secretary of the Air Force and as director of operations at the Air Education and Training Command. Following the events of 9/11, he served as a director of the Air Force Crisis Action Team in the Pentagon. General Johnson is a member of several organizations, including the Air Force Association, the Military Officers Association of America, Military Order of World Wars, Veterans of Foreign Wars, Reserve Officers Association, League of United Latin American Citizens, Women in Aviation, the International Black Aerospace Council, Inc., and Tuskegee Airmen, Inc. General Johnson was elected to a 2-year term as the Tuskegee Airmen, Inc., national president in July 2010. He recently retired from United Parcel Service (UPS) after nearly 20 years of service, where he served as the flight operations employment manager and concluded his career working on a special project as the manager of airline manuals. In 2011, General Johnson was awarded a doctorate in humane letters by Tuskegee University. In November 2011, he received an appointment by the Secretary of the Air Force to the Civil Air Patrol Board of Governors, the senior policy-making body for that body as established by Public Law. He is a trustee of the U.S. Air Force Falcon Foundation and is a member of the NSB.

Catherine M. Kelleher is professor for public policy at the University of Maryland and senior faculty associate at Brown University's Watson Institute, where her research interests include cooperative European defense and security policies, North Atlantic Treaty Organization relations, and international security and arms control. Dr. Kelleher served in the Clinton administration as personal representative of the Secretary of Defense in Europe and as deputy assistant secretary of defense for Russia, Ukraine, and Eurasia. She has served on numerous scientific boards and advisory committees, including as a member of the NRC Committee on National Security Implications of Climate Change for U.S. Naval Forces and the NRC Committee on the "1,000-Ship Navy"—A Distributed and Global Maritime Network. She is a former member of the NSB.

Jeffrey E. Kline is a professor of practice in the Operations Research Department and program director, Consortium for Robotics and Unmanned Systems Education and Research (CRUSER), at the Naval Postgraduate School (NPGS). He oversees over 25 interagency and interschool research and educational initiatives related to maritime security, maritime domain awareness, port security, counterpiracy operations, and maritime critical infrastructure, with sponsors ranging from the Secretary of the Navy, Office of the Secretary of Defense, Department of Energy, Secretary of the Navy, and the U.S. Coast Guard. He retired as a captain from the U.S. Navy and has over 26 years of extensive naval operational experience, including commanding two U.S. Navy ships and serving as deputy of operations for the Commander, Sixth Fleet, where he participated in theaterwide operational planning. In addition to his sea service, Mr. Kline spent 3 years as a naval analyst in the Office of the Secretary of Defense. He is a 1992 graduate of the NPGS's Operations Research Program, where he earned the Chief of Naval Operations Award for Excellence in Operations Research, and is a 1997 distinguished graduate of the National War College, where he earned the Chairman of the Joint Chief's Strategic Writing Award. Mr. Kline's NPGS faculty awards include the 2011 Institute for Operations Research and Management Science Award for Teaching of Practice, the 2007 Hamming Award for interdisciplinary research, the 2007 Wayne E. Meyers Award for Excellence in Systems Engineering Research, and the 2005 Northrop Grumman Award for Excellence in Systems Engineering.

Annette J. Krygiel is currently an independent consultant with expertise in the management of large-scale systems, particularly in regard to software development and systems integration. She served as a distinguished visiting fellow at the Institute for National Strategic Studies at the National Defense University (NDU), where she wrote a book on large-scale system integration. Prior to that, she was director of the Central Imagery Office (CIO), a Department of Defense combat support agency, until CIO joined the National Imagery and Mapping Agency in October 1996. Dr. Krygiel began her career at the Defense Mapping Agency, where she held various positions, including chief scientist. Dr. Krygiel previously served as chair of the NRC Committee on the Role of Experimentation in Building Future Naval Forces and recently served as a member of the NRC Committee on U.S. Naval Forces' Capabilities for Responding to Small Vessel Threats.

Thomas V. McNamara is currently director, strategy and business creation at Raytheon Integrated Defense Systems. He served previously as senior vice president and chief technology officer for Textron Systems, where he focused on long-term strategic technical investments and program execution to support Textron Systems' businesses. He was responsible for the development of technology and systems to address the emerging challenges in the areas of precision engagement, maritime and land platforms, advanced controls, and aircraft engines. His areas of expertise include guidance, navigation and control; intelligent autonomy; precision weapons delivery; micro-electromechanical sensors; dismounted soldier systems; mission planning; and systems integration for naval submersible and aircraft platforms. He recently served as a member of the NRC Committee on the "1,000-Ship Navy"—A Distributed and Global Maritime Network, the Committee on Distributed Remote Sensing for Naval Undersea Warfare, and as co-chair of the Committee on U.S. Naval Forces' Capabilities for Responding to Small Vessel Threats.

Richard W. Mies, ADM, USN (Retired), is the CEO and president of the Mies Group, Ltd. He provides strategic planning and risk assessment advice and assistance to clients on international security, energy, defense, and maritime issues. A distinguished graduate of the U.S. Naval Academy, he completed a 35-year career as a nuclear submariner in the U.S. Navy and commanded the U.S. Strategic Command for 4 years prior to retirement in 2002. Admiral Mies served as a senior vice president and deputy group president of Science Applications International Corporation (SAIC) and as the president and chief executive officer of Hicks and Associates, Inc., a wholly owned subsidiary of SAIC from 2002 to 2007. He also served as the chairman of the Department of Defense Threat Reduction Advisory Committee from 2004 to 2010 and as the chairman of the board of the Navy Mutual Aid Association from 2003 to 2011. He presently serves as chairman of the Strategic Advisory Group of the U.S. Strategic Command and chairman of the board of the Naval Submarine League; more recently, he became a trustee of the U.S. Naval Academy Foundation. He is a member of the Committee on International Security and Arms Control of the NAS, a member of the boards of governors of Los Alamos National Laboratory and Lawrence Livermore National Laboratory, and a member of the board of directors of Mutual of Omaha Company, Babcock and Wilcox Company, and Exelon Corporation. He also serves on numerous advisory boards. Admiral Mies completed postgraduate education at Oxford University, England, the Fletcher School of Law and Diplomacy, and Harvard University. He holds a master's degree in government administration and international relations.

C. Kumar N. Patel (NAS/NAE) is the founder, president, and CEO of Pranalytica, Inc., a Santa Monica-based company that is the leader in quantum cascade laser technology for defense and homeland security applications. He is also professor of physics and astronomy, electrical engineering, and chemistry at the University of California, Los Angeles (UCLA). He served as vice chancellor for research at UCLA from 1993 to 1999. Prior to joining UCLA, he was the executive director of the Research, Materials Science, Engineering and Academic Affairs Division at AT&T Bell Laboratories, where he began his career by carrying out research in gas lasers. He is the inventor of the carbon dioxide laser and many other molecular gas lasers that ushered in the era of high-power sources of coherent optical radiation. Dr. Patel was awarded the National Medal of Science for his invention of the carbon dioxide laser. His other awards include the Ballantine

Medal of the Franklin Institute, the Zworykin Award of the National Academy of Engineering, the Lamme Medal of the IEEE, the Texas Instruments Foundation Founders Prize, and many more. Dr. Patel holds a B.E. in telecommunications from the College of Engineering in Poona, India, and received his M.S. and Ph.D. in electrical engineering from Stanford University.

Heidi C. Perry is director, algorithms and software, at the Charles S. Draper Laboratory, Inc. Previously she was director, internal research and development, at the Draper Laboratory. Her expertise includes guidance, navigation, and control; global position system antijam and ground control; precisions weapons delivery command and control; autonomous systems; mission-critical software; and C4ISR systems. She served as a member of the NRC Committee on National Security Implications of Climate Change for U.S. Naval Forces and on the NRC Committee on the "1,000-Ship Navy"—A Distributed and Global Maritime Network. She is a member of the NSB.

Gene H. Porter is an adjunct staff member at the Institute for Defense Analyses. His areas of expertise include national security planning and weapons systems development and defining the defense planning scenarios that are intended to guide the development of the U.S. military force structure. Mr. Porter formerly served as the director of acquisition policy and program integration at the Office of the Under Secretary of Defense for Acquisition. He has served on numerous scientific boards and advisory committees, including as chair of NRC Committee for Mine Warfare Assessment and more recently as a member of the NRC Committee on U.S. Naval Forces' Capabilities for Responding to Small Vessel Threats.

Dana R. Potts is the senior Navy experienced systems engineer principal for the Horizontal Integration Operational Concepts Team at Lockheed Martin Aeronautics Advanced Development Programs (Skunk Works), having retired from the U.S. Navy with the rank of captain. He is responsible for projects that include any maritime component. He was scenario lead for two corporate-level Marine Air Ground Task Force Experiments, creating the scenarios and coordinating the efforts involving aeronautics, electronic systems, enterprise operations, information systems and global solutions, mission systems, and space systems. He also led the man-in-the-loop experiment studying the attributes of the Skunk Works concept for the Unmanned Carrier Launched Airborne Surveillance and Strike system. He has received two Lockheed Martin Nova awards for projects involving teamwork with these scenarios. Prior to joining Lockheed, Mr. Potts completed over 28 years with the U.S. Navy and held leadership positions both ashore and at sea, primarily in tactical aviation flying the F-4 Phantom and the F-14 Tomcat. His significant operational experience included command of Fighter Squadron ONE FIVE FOUR and command of Carrier Air Wing SEVENTEEN in combat operations. He was also a fellow with the CNO Strategic Studies Group XXIII that conducted research, developed innovative concepts, and made recommendations to the Chief of Naval Operations concerning the "Navy After Next." He earned a B.S. in computer science from Texas A&M and an M.S. in national security strategy from the NDU National War College.

John E. Rhodes, LtGen, USMC (Retired), is currently an independent consultant having retired from the U.S. Marine Corps with the rank of lieutenant general after 36 years of service. His background is in development of warfighting concepts and in the integration of all aspects of doctrine, organization, training and education, equipment, and support and facilities to enable the Marine Corps to field combat-ready forces. In his last position, General Rhodes served as commanding general of the Marine Corps Combat Development Command, where his responsibilities included assessments of current and future operating environments and adaptation of the Corps' training infrastructure and resources in order to ensure that integrated capabilities were delivered to the combatant commanders. General Rhodes has served on numerous scientific boards and advisory committees, including as a member of the NRC Committee on Manpower and Personnel Needs for a Transformed Naval Force and the Committee on U.S. Forces' Capabilities for Responding to Small Vessel Threats; he is a member of the NSB.

Robert M. Stein is currently an independent consultant, having served previously as vice president of the Raytheon Company until he retired in 2000. He managed Raytheon's Advanced Systems Office. He was responsible for the formulation and implementation of advanced systems and concepts for current and future Raytheon product lines. Mr. Stein led concept formulation and advanced development studies for the company and the U.S. government, addressing the advanced strategic and tactical defense needs for the United States and many of its allies. These have ranged from early concept studies on the protection of the continental United States (CONUS) and the defense of retaliatory forces against nuclear attack in the 1960s, to tactical defense of land and sea forces in the 1970s, to defense of CONUS, theater, and allied military and civilian assets against air, cruise, or ballistic missile attack in the 1980s and 1990s. He has participated in a number of Army Science Board, Air Force Scientific Advisory Board, and Navy Research Advisory Committee task forces. He has served on and/or cochaired many DSB task forces and summer studies and is currently a senior fellow on the board. He also currently serves as a member of the Missile Defense Agency Advisory Committee. Mr. Stein performed undergraduate work in electrical engineering at MIT and has performed extensive graduate studies at MIT and Boston University in mathematical physics. He holds a patent in multibeam radar antenna techniques, has published numerous articles on defense technology and related policy issues, and has taught a variety of courses on radar and information theory.

In 1992, Raytheon awarded Mr. Stein the Thomas L. Phillips Award of Excellence in Technology—the company's highest recognition for technical achievement.

Vincent Vitto is the retired president and CEO of Charles Stark Draper Laboratory, Inc., where he served for 9 years until 2006. Since 2006, he has been working as an independent consultant. Before joining Draper in 1997, he spent 32 years at MIT's Lincoln Laboratory, rising to assistant director of surface surveillance and communications. He holds an M.S. in physics from Northeastern University and a B.S. in physics from the Polytechnic Institute of Brooklyn. Currently, Mr. Vitto is chairman of the board of directors of Mercury Computer Systems, a member of the QinetiQ North America proxy board of Directors and a member of the board of trustees for the Aerospace Corporation. He is also a member of the National Associates of the National Academies and a fellow of the AIAA. He serves on the board of trustees at the Massachusetts Eye and Ear Infirmary. Mr. Vitto has received numerous public service awards in his career, including the Meritorious Public Service Award and the Superior Public Service Award from the Department of the Navy, the Decoration for Exceptional Civilian Services from the Department of the Air Force, and, in 2009, the Department of Defense Medal for Distinguished Public Service Award. He has served on numerous scientific boards and advisory committees, including the Intelligence Science Board, the DSB, NRO's Technical Advisory Group, and the National Geospatial Intelligence Agency Advisory Group. He also chaired the NSB from 1999 to 2004.

David A. Whelan (NAE) is vice president, Engineering, for Boeing Defense Space & and Security (BDS). Dr. Whelan has broad responsibility to create, seek out, and explore new technology and growth vectors for the Boeing Company. Boeing's technology and systems span a wide range of government missions, from space and airborne systems to ground systems to undersea systems. He has in-depth knowledge of science, technology, systems, and future customer requirements, enabling Boeing to find new solutions to world's most challenging problems. Dr. Whelan serves as a member of the Technical Advisory Committee for HRL Laboratories, the legacy R&D laboratory of the former Hughes Aircraft Company, a LLC jointly owned by Boeing and GM. Prior assignments include vice president for strategy and innovation and chief scientist, BDS; vice president, Boeing corporate business development and strategy; and vice president/general manager and deputy to the president of the Boeing Company. He began his career with Boeing as vice president and chief technology officer for the space and communications group. Dr. Whelan is a fellow of the American Physical Society and of the American Institute of Aeronautics and Astronautics and a senior member of the Institute of Electrical and Electronic Engineers. He has numerous publications on electromagnetic radiation, laser plasma phenomena, and defense systems. He holds over 150 patents on navigation systems, radar systems, antenna, and low-observable technology. He is the recipient of the Secretary of Defense medals for Outstanding Public Service (1998) and Meritorious Civil Service (2001). Before joining Boeing, Dr. Whelan served as director of the Tactical Technology Office (SES-5) of the Defense Advanced Research Projects Agency (DARPA). During his early career, he worked at Northrop where he was one of the key designers of the B-2 stealth bomber and contributed to the YF-23 advanced tactical fighter, at Hughes Aircraft Company, and at the Department of Energy's Lawrence Livermore National Laboratory (LLNL).

Peter G. Wilhelm (NAE) is director of the Naval Center for Space Technology (NCST) at the NRL. He is responsible for the technical and managerial leadership of NCST's mission, which is to preserve and enhance a strong space technology base and provide expert assistance in the development and acquisition of space systems that support naval missions. During Mr. Wilhelm's tenure, the space program at NRL has grown from a branch to a division of the center. Under his direction, NCST and the Navy have achieved numerous successes and firsts in space, including the GPS satellite and the highly successful Clementine Deep Space Mission, which demonstrated the capability of low-cost, high-value space exploration and has become the model for it. Mr. Wilhelm's achievements include contributions to the design, development, and operation of 100 scientific and fleet-support satellites. Mr. Wilhelm is a fellow of the AIAA and of the Washington Academy of Science.

John D. Wilkinson is an assistant group leader at MIT Lincoln Laboratory, in the Air Defense Techniques Group of the Air and Missile Defense Technology Division. He has worked at Lincoln Laboratory since 1998, beginning in the Intelligence, Test, and Evaluation Group. After a decade of radar data analysis, radar system engineering, and radar testing experience, Mr. Wilkinson now serves as the Lincoln Program Manager for several science and technology programs related to air defense. In this role he proposed and led the development of an UHF radar installed on Lincoln's Boeing 707 and helped design, build, and deploy two other radar systems as well. He was awarded a B.S. in physics from the University of Massachusetts at Amherst and an M.S. in electrical engineering from Tufts University.

Staff

Charles F. Draper is director of the National Research Council's Naval Studies Board (NSB). He joined the NSB in 1997 as program officer then senior program officer and in 2003 became associate director and acting director of the NSB. During his tenure with the NSB, Dr. Draper has served as study director on a wide range of topics aimed at helping the Department of the Navy and DOD with their scientific, technical, and strategic planning. He served as study director for the report *Conventional Prompt Global Strike: Issues for 2008 and Beyond* and the more recent *Making Sense of Ballistic Missile Defense: An Assessment of Concepts and Systems for U.S. Boost-Phase Missile Defense in Comparison to Other Alternatives.* Before joining the NSB, Dr. Draper was the lead mechanical engineer at S.T. Research Corporation, where he provided technical and program management support for satellite Earth station and small satellite design. He received his Ph.D. in mechanical engineering from Vanderbilt University in 1995; his doctoral research was conducted at the Naval Research Laboratory (NRL), where he used an atomic-force microscope to measure the nanomechanical properties of thin-film materials. In parallel with his graduate student duties, Dr. Draper was a mechanical engineer with Geo-Centers, Inc., working on-site at NRL on the development of an underwater X-ray backscattering tomography system used for the nondestructive evaluation of U.S. Navy sonar domes on surface ships.

Douglas C. Friedman is a program officer with the Board on Chemical Sciences and Technology at the National Research Council of the National Academy of Sciences. His primary scientific interests lie in the fields of organic and bio-organic materials and chemical and biological sensing and nanotechnology, particularly as they apply to national and homeland security. Dr. Friedman has supported a diverse array of activities since joining the NRC. He has directed studies in the areas of carbohydrate chemistry and glycobiology, crude oil pipeline transportation, computational molecular dynamics simulations, and chemical and biological defense. Dr. Friedman has also supported activities in biomass utilization, critical resources, and antibiotics research and development. Prior to joining the NRC, Dr. Friedman performed research in physical organic chemistry and chemical biology at Northwestern University, the University of California, Los Angeles, the University of California, Berkeley, and Solulink Biosciences. He holds a Ph.D. in chemistry from Northwestern University and a bachelor's degree in chemical biology from the University of California, Berkeley.

D

Acronyms and Abbreviations

A2A	air-to-air (weapon)
A2/AD	antiaccess/area denial
ALCM	air-launched cruise missile
AOR	area of responsibility
ARCI	acoustic rapid COTS insertion
ASN RDA	Assistant Secretary of the Navy for Research, Development, and Acquisition
ASUW	antisurface warfare
ASW	antisubmarine warfare
B-52	long-range, subsonic, jet-powered strategic bomber (Boeing)
BMD	ballistic missile defense
C2	command and control
C3	command, control, and communications
CANES	Consolidated Afloat Networks and Enterprise Services
CBRN	chemical, biological, radiation, or nuclear
CCG (also CUSCG)	Commandant U.S. Coast Guard
CEC	cooperative engagement capability
CG	cruiser
CLF	combat logistics force
CLS	contractor logistics supplied
CMC	Commandant, Marine Corps

CNA	Center for Naval Analyses
CNAS	Center for a New American Security
CNE	Commander, U.S. Naval Forces, Europe
CNO	Chief of Naval Operations
CNR	Chief of Naval Research
COCOM	combatant commander
COMPACFLT	Commander, Pacific Fleet
CONOPS	concept of operations
CONUS	continental United States
COP	common operational picture
COTS	commercial off-the-shelf
CPF	Commander, U.S. Pacific Fleet
CSG	carrier strike group
CTO	chief technology officer
CVN	nuclear-powered aircraft carrier
CWID	Coalition Warrior Interoperability Demonstration
DARPA	Defense Advanced Research Projects Agency
DAS	Defense Acquisition System
DASN RDT&E	Deputy Assistant Secretary of the Navy for Research, Development, Testing and Evaluation
DDG	guided missile destroyer
DFARS	Defense Federal Acquisition Regulation Supplement
DICE-T	Dismounted Interactive Counter-IED Environment for Training
DOD	Department of Defense
DOTMLPF	doctrine, organization, training, material, leadership and education, personnel, and facilities
DRRS	Defense Readiness Reporting System
DSB	Defense Science Board
EMCON	emission control
ESM	electronic support measure
EW	electronic warfare
F/A	fight/attack (aircraft)
FARS	Federal Acquisition Regulations System
FAV	fast attack vehicle
FFC	Fleet Forces Command
FFRDC	Federally Funded Research and Development Center
FLEX	fleet experimentation
FORCECOM	Force Readiness Command (U.S. Coast Guard)

GMM	gun mission module (on an LCS)
GPS	Global Positioning System
HA/DR	humanitarian aid and disaster relief
HSCB	human social, cultural, and behavioral
I2	integration and interoperability
IC	intelligence community
ICD	interface control documents
ICT	Institute for Creative Technologies
IED	improvised explosive device
INOTS	Immersive Naval Officer Training System
IOC	initial operating capability
IPT	integrated process team
ISO	International Organization for Standardization
ISR	intelligence, surveillance, and reconnaissance
JCIDS	Joint Capabilities Integration and Development System
JFEX	Joint Expeditionary Force Exercise
JHU/APL	Johns Hopkins University/Applied Physics Laboratory
JIEDDO	Joint Improvised Explosive Device Defeat Organization
JROC	Joint Requirements Oversight Council
JTRS	Joint Tactical Radio System
JUONS	Joint Urgent Operational Needs Statement
LANTFLT	Atlantic Fleet
LCS	littoral combat ship
LHD	amphibious assault ship (multipurpose)
LOP	local operational picture
LRU	life raft unit (one man, V-bottom inflatable life raft)
LTPA	long-term pricing agreement
MCCDC	Marine Corps Combat Development Command
MCM	mine countermeasures
MCWL	Marine Corps Warfighting Laboratory
MDA	Military Designate Authority; Missile Defense Agency
MEMS	microelectromechanical systems
MET	mission essential task
MILES	Multiple Integrated Laser Engagement System
MOE	measure of effectiveness
MOOTW	military operations other than war
MRAP	mine-resistant ambush-protected (vehicle)
M&S	modeling and simulation

MSM	maritime security module (on an LCS)
N1	Manpower, Personnel, Training, and Education
N2/N6	Information Dominance
N3/N5	Operations, Plans, and Strategy
N4	Fleet Readiness and Logistics
N8	Integration of Capabilities and Resources
N81	Assessment Division
N9	Warfare Systems
NGO	nongovernmental organization
NIFC-CA	Naval Integrated Fire Control
NLOS	non-line-of-sight
NRAC	Naval Research Advisory Committee
NRC	National Research Council
NRL	Naval Research Laboratory
NRO	National Reconnaissance Office
NRRE	Navy Readiness Reporting Enterprise
NRRS	Navy Readiness Reporting System
NSAWC	Naval Strike and Air Warfare Center
NSB	Naval Studies Board
NWDC	Navy Warfare Development Command
OCO	overseas contingency operation
OCONUS	outside continental United States
OFP	operational flight program
O&M	operations and maintenance
ONI	Office of Naval Intelligence
ONR	Office of Naval Research
ONR-G	Office of Naval Research-Global
OODA	observe, orient, decide, act
OPNAV	Office of the Chief of Naval Operations
OPSEC	operations security
OSD	Office of the Secretary of Defense
OSD AT&L	Office of the Secretary of Defense for Acquisition, Technology, and Logistics
OSD RDT&E	Office of the Secretary of Defense for Research, Development, Test, and Evaluation
OTA	other transaction authority
OT&E	operational test and evaluation
OUSD/AT&L	Office of the Under Secretary of Defense for Acquisition, Technology, and Logistics
PACFLT	Pacific Fleet

PACOM	Pacific Command
PC	personal computer
PEO	Program Executive Office
PESTO	personnel, equipment, supply, training, and ordnance
PM	program/project manager
PN&T	positioning, navigation, and timing
POM	program objective memorandum
PPBE	programming, planning, budgeting, and execution
PSO	peace support operation
QRC	quick reaction capability
RAFA	Rapid Acquisition and Fielding Agency
RCS	radar cross section
R&D	research and development
RF/IF	radio frequency/intermediate frequency
RIM-116 RAM	RIM-116 Rolling Airframe Missile
RIMPAC	Rim of the Pacific
ROEs	rules of engagement
ROV	remotely operated vehicle
SBIR	Small Business Innovation Research
SIG	senior integration group
SM	Standard Missile
SME	subject matter expert
SOCOM	Special Operations Command
SOF	Special Operations Forces
SOLAS	safety of life at sea
SORTS	Status of Resources and Training System
SSBN	nuclear-powered, ballistic missile submarine
SSGN	nuclear-powered, guided missile submarine
S&T	science and technology
SUW	surface warfare
SYSCOM	Systems Command
TM	technical manual
TPM	technical performance measure
TSG	threat study group
TTPs	tactics, techniques, and procedures
UARC	university affiliated research center
USA	U.S. Army
USCG	U.S. Coast Guard

USCYBERCOM	U.S. Cyber Command
USG	U.S. government
USMC	U.S. Marine Corps
USN	U.S. Navy
VCNO	Vice Chief of Naval Operations
VLS	vertical-launch system
WCB	warfighting capability baseline

E

Glossary

Brevity code: Code that provides no security but which has as its sole purpose the shortening of messages rather than the concealment of their content.

Commander's Guidance: Written by the commander for his command and promulgated for all to read.

Commander's Intent: Written statement by the commander to his subordinate commanders and staff which is a personal expression of the purpose of the operation.

Common operational picture: Single identical display of relevant information shared by more than one command.

Concept of operations: General description of actions to be taken in pursuit of mission accomplishment.

Cyber operation or cyberspace operations: Employment of cyberspace capabilities primarily to achieve objectives in or through cyberspace. Such operations include computer [network operations] and activities to operate and defend DOD information networks.

NOTE: Based on Department of Defense, 2010, *Department of Defense Dictionary of Military and Associated Terms*, Joint Publication 1-02, November 8 (as amended through December 15, 2012).

Deception event: Event executed at a specific time and location in support of a deception operation.

Disinformation: Information that is intentionally false.

Emission control: Selective and controlled use of electromagnetic, acoustic, or other emitters to optimize command and control capabilities while minimizing, for operations security, the following: (a) detection by enemy sensors; (b) mutual interference among friendly systems; and/or (c) enemy interference with the ability to execute a military deception plan.

Information operations: Integrated employment, during military operations, of information-related capabilities in concert with other lines of operation to influence, disrupt, corrupt, or usurp the decision making of adversaries and potential adversaries while protecting our own.

Irregular warfare: Violent struggle among state and nonstate actors for legitimacy and influence over the relevant population(s). Irregular warfare favors indirect and asymmetric approaches, though it may employ the full range of military and other capacities, in order to erode an adversary's power, influence, and will.

Limited access: Plan, operation, or mission that has access restricted to specific individuals, units, or commands.

Local operational picture: Display that permits all friendly forces to have the same information and situational awareness at the local unit level.

Misinformation: Information that is unintentionally false.

Military operations other than war: Encompasses the use of military capabilities across the range of military operations short of war. These military actions can be applied to complement any combination of the other instruments of national power and occur before, during, and after war.

Operations security: Process of identifying critical information and subsequently analyzing friendly actions attendant to military operations and other activities.

Procedures: Standard, detailed steps that prescribe how to perform specific tasks.

Rules of engagement: Directives issued by competent military authority that delineate the circumstances and limitations under which U.S. forces will initiate and/or continue combat engagement with other forces encountered.

Tactics: Employment and ordered arrangement of forces in relation to each other

Techniques: Nonprescriptive ways or methods used to perform missions, functions, or tasks.

F

Study Briefings and Organizational Interfaces

Meeting Number 1 (Inaugural Meeting), February 28-March 1, 2012, Keck Center of the National Academies, 500 5th Street, NW, Washington, D.C.

- Information and intelligence perspectives: chief scientist, Office of Naval Intelligence Scientific and Technical Center.
- Maritime intelligence perspectives: deputy director, National Maritime Intelligence Center.
- Operational perspectives: assistant deputy chief of naval operations for Operations, Plans, and Strategy; Office of the Chief of Naval Operations (OPNAV), N3/N5.
- U.S. Marine Corps perspectives: commanding general, Marine Corps Combat Development Command, and deputy commandant for combat development and integration, U.S. Marine Corps.
- Office of Naval Research perspectives: executive director, Office of Naval Research.
- Fleet readiness and logistics perspectives: deputy chief of naval operations for fleet readiness and logistics, OPVAV N4.
- Cyberperspectives: deputy commander, U.S. Fleet Cyber Command, deputy commander, U.S. TENTH Fleet.
- Rapid prototyping perspectives: assistant commandant for capability, U.S. Coast Guard.
- Defense Advanced Research Projects Agency (DARPA) perspectives: program manager, Tactical Technology Office, DARPA.

Meeting Number 2, April 11-12, 2012, Keck Center of the National Academies.

- OPNAV N81 perspectives: deputy director, Assessment Division, OPNAV N81.
- ASN (RD&A) perspectives: deputy assistant secretary of the Navy for Research, Development, Test and Evaluation.
- Former commandant, U.S. Coast Guard perspectives: former commandant, USCG.
- Program Executive Office (PEO) Littoral Combat Ship perspectives: PEO, Littoral Combat Ships.
- Naval War College perspectives: professor, Analysis Department, Naval War College.
- National War College perspectives: professor of international history, National War College.
- SSBN Security Program perspectives: SSBN security technical director, OPNAV N97; SSN/SSGN survivability technical director, OPNAV N97; head, Submarine Posture, OPNAV N97.
- Chief of Naval Operations (CNO) Strategic Studies Group perspectives: deputy director, CNO Strategic Studies Group.
- National Research Council Committee on Avoiding Technology Surprise for Tomorrow's Warfighter perspectives: senior system engineer, The Tauri Group, and chair and study director, Committee on Avoiding Technology Surprise for Tomorrow's Warfighter.

Subgroup Site Visit, May 2, 2012, Tactical Electronic Warfare Division, Naval Research Laboratory, 4555 Overlook Avenue, SW, Washington, DC 20375.

- Naval tactical electronic warfare capabilities and research: superintendent, Tactical Electronic Warfare Division, Naval Research Laboratory.

Meeting Number 3, May 16-17, 2012, Keck Center of the National Academies.

- NeXTech perspectives: director, Emerging Capabilities Division, Rapid Reaction Technology Office, Deputy Assistant Secretary of Defense (Rapid Fielding).
- Air Force red team—Lincoln Laboratory Air Vehicle Survivability Evaluation Program: assistant group leader, Systems and Analysis Group, MIT Lincoln Laboratory.
- Space systems perspectives—A day without space: deputy director, Strategic and Space Systems (S&SS), Office of the Assistant Secretary of Defense for Research and Engineering (OASD/R&E) and Office of the Under Secretary of Defense for Acquisition, Technology, and Logistics (OUSD/AT&L); senior ana-

lyst, Office of the Deputy Chief of Naval Operations for Information Dominance, OPNAV N2/N6.

- Operation Burnt Frost perspectives: program executive, Aegis Ballistic Missile Defense, Missile Defense Agency; department head, Air and Missile Defense, Johns Hopkins University, Applied Physics Laboratory.
- Navy Warfare Development Command (NWDC) perspectives: Assistant Chief of Staff for Concepts, NWDC.
- Pacific Fleet perspectives: director, warfighting assessment and readiness, Commander, Pacific Fleet.

Meeting Number 4, June 26-28, 2012, Naval Surface Warfare Center (NSWC), Corona Division, Corona, Calif. (June 26); Arnold and Mabel Beckman Center, Irvine, Calif. (June 27-28).

- NSWC perspectives on capability surprise: commanding officer, NSWC Corona Division.
- Space and Naval Warfare Systems Command (SPAWAR) perspectives on capability surprise and "A day without space": science and technology competency lead; head, research and applied sciences, SPAWAR Systems Center Pacific; head, research and applied sciences, SPAWAR Systems Center Pacific; head, space systems engineering, SPAWAR Systems Center Pacific.
- I Marine Expeditionary Force (I MEF) perspectives on capability surprise: deputy commanding general, I MEF, and commanding general, 1st Marine Expeditionary Brigade.
- U.S. Pacific Fleet (PACFLT) perspectives on capability surprise: director, warfighting and readiness, PACFLT.
- Naval Special Warfare Command (NAVSPECWARCOM) perspectives on capability surprise: commander, NAVSPECWARCOM.
- U.S. Coast Guard perspectives on capability surprise: chief of response, Pacific area, PAC-33, U.S. Coast Guard.
- Naval Mine and Anti-Submarine Warfare Command (NMAWC) perspectives: deputy commander, NMAWC.
- Academia perspectives: professor and Roger C. Lipitz Chair and Director, Center for Public Policy and Private Enterprise, School of Public Policy, University of Maryland, College Park.
- Commercial perspectives: executive director, Institute for Creative Technologies.
- Commercial perspectives: senior staff member, Google.
- Commercial perspectives: fellow and chief technology officer, Global Application Innovation Services, IBM.

Meeting Number 5, July 26-27, 2012, Keck Center of the National Academies.

- Intelligence community perspectives: former top-level federal intelligence officials; research staff member, Intelligence Analysis Division, Institute for Defense Analyses.
- Intelligence Advanced Research Projects Agency (IARPA) perspectives: acting director, IARPA.
- Operational Test and Evaluation Force (OPTEVFOR) perspectives: commander, OPTEVFOR.
- U.S. Air Force and special mission aircraft perspectives: acting deputy director, "Big Safari Program," U.S. Air Force.
- U.S. Army perspectives: military deputy to chief scientist, U.S. Army.
- U.S. Marine Corps perspectives on red teaming: acting director (G-3/5) for deputy commandant for combat development and integration; U.S. Marine Corps red team.
- Office of Naval Research (ONR) perspectives: commanding officer, ONR Global; technical director, ONR Global.

Subgroup Site Visit, August 21, 2012, Fleet Forces Command, 1562 Mitscher Avenue, Norfolk, Va.

- U.S. NWDC perspectives: chief of staff, NWDC.
- U.S. Fleet Forces Command perspectives: deputy commander for fleet management and chief of staff, N03A, U.S. Fleet Forces Command.